Information and Instructions

T0261759

This shop manual contains several sections each covering a specific group of wheel type tractors. The Tab Index on the preceding page can be used to locate the section pertaining to each group of tractors. Each section contains the necessary specifications and the brief but terse procedural data needed by a mechanic when repairing a tractor on which he has had no previous actual experience.

Within each section, the material is arranged in a systematic order beginning with an index which is followed immediately by a Table of Condensed Service Specifications. These specifications include dimensions, fits, clearances and timing instructions. Next in order of arrangement is the procedures paragraphs.

In the procedures paragraphs, the order of presentation starts with the front axle system and steering and proceeding toward the rear axle. The last paragraphs are devoted to the power take-off and power lift systems. Interspersed where needed are additional tabular specifications pertaining to wear limits, torquing, etc.

HOW TO USE THE INDEX

Suppose you want to know the procedure for R&R (remove and reinstall) of the engine camshaft. Your first step is to look in the index under the main heading of ENGINE until you find the entry "Camshaft." Now read to the right where under the column covering the tractor you are repairing, you will find a number which indicates the beginning paragraph pertaining to the camshaft. To locate this wanted paragraph in the manual, turn the pages until the running index appearing on the top outside corner of each page contains the number you are seeking. In this paragraph you will find the information concerning the removal of the camshaft.

More information available at haynes.com
Phone: 805-498-6703

J H Haynes & Co. Ltd.
Haynes North America, Inc.

ISBN-10: 0-87288-072-9
ISBN-13: 978-0-87288-072-6

SHOP MANUAL

JOHN DEERE

SERIES 520-530-620-630-720-730

(Gasoline, All-Fuel and LP-Gas)

IDENTIFICATION

Tractor serial number stamped on plate on right side of main case.

INDEX (By Starting Paragraph)

	Series 520 & 530	Series 620 & 630	Series 720 & 730		Series 520 & 530	Series 620 & 630	Series 720 & 730
BELT PULLEY	120	120	120	FRONT SYSTEM (Tricycle)			
BRAKES	150	150	150	Pedestal	1	1	1
CARBURETOR (Not LP-Gas)	68	68	68	"Roll-O-Matic"	4	4	4
CARBURETOR (LP-Gas)	87	87	87	Vertical spindle	1	1	1
COOLING SYSTEM				FRONT SYSTEM (Axle)			
Fan and shaft	107	107	107	Axle pivot pins	7	7	7
Radiator	105	105	106	Drag links	6	6	6
Thermostat	109	109	109	Pedestal	8	8	8
Water pump	111	111	111	Steering knuckles	5	5	5
DIFFERENTIAL				Vertical spindle	8	8	8
Overhaul	142	144	144	GOVERNOR			
Remove and reinstall	141	143	143	Adjust	100	100	100
ENGINE				Overhaul	101	101	101
Cam followers	48	49	49	Reclaiming case	104	104	104
Camshaft and bearings	45	46	46	"POWR-TROL"			
Connecting rods and bearings	53	53	53	Dual remote cyl. valve housing	194	194	194
Crankcase cover	61	61	61	Lubrication and bleeding	176	176	176
Crankshaft	58	58	58	Operating adjustments	178	178	178
Cylinder block	54	54	54	Pump	191	191	191
Cylinder head	30	30	31	Remote cylinder	203	203	203
Flywheel	59	59	59	Rockshaft hsg. and components	195	195	195
Ignition timing	113	113	113	Rockshaft relief valve housing	192	192	192
Main bearings	55	55	55	Service tests and adjustments	181A	181A	181A
Oil filter	63	63	63	Single remote cyl. valve hsg.	193	193	193
Oil pressure	62	62A	62A	Trouble-shooting	177	177	177
Oil pump	65	65	65	POWER TAKE-OFF			
Pistons and rings	51	51	51	Adjust clutch	165	165	165
Piston pins	52	52	52	Clutch overhaul	167	167	167
Piston and rod removal	50	50	50	Bevel gears	169	169	169
Oil seal (left hand)	57	57	57	Oil pump	168	168	168
Oil seal (right hand)	56	56	56	Output shaft and seal	167	167	167
Rocker arms	38	38	38	REAR AXLE	147	147, 148	147, 148
Timing gears	40	40	40	STEERING SYSTEM			
Valves and seats	32	32	32	Manual	9	9, 15A	9
Valve guides and springs	36	36	36	Power	16	16, 29C	16
Valve rotators	35	35	35	TRANSMISSION			
Valve timing	39	39	39	Countershaft	128	135	140A
FINAL DRIVE (Except "Hi-Crop")				Drive shaft	127	134	140
Axle shafts	147	147	147	Reduction gear cover	123	130	136
Bull gears—R&R	147	147	147	Shifters and shafts	125	132	138
Bull pinions—R&R	142	144	144	Sliding gear shaft	126	133	139
FINAL DRIVE ("Hi-Crop")				Top cover	124	131	137
Axle shafts	....	148	148				
Bull gears—R&R	....	148	148				
Chains and sprockets	....	148A	148A				
Drive shaft and pinion	....	149A	149A				

CONDENSED SERVICE DATA

GENERAL	Series 520 & 530 Gasoline	All-Fuel	LP-Gas	Series 620 & 630 Gasoline	All-Fuel	LP-Gas	Series 720 & 730 Gasoline	All-Fuel	LP-Gas
Engine Make	Own	Own	Own	Own	Own	Own	Own	Own	Own
Number of Cylinders	2	2	2	2	2	2	2	2	2
Bore—Inches	4 11/16	4 11/16	4 11/16	5 1/2	5 1/2	5 1/2	6	6	6
Stroke—Inches	5 1/2	5 1/2	5 1/2	6 3/8	6 3/8	6 3/8	6 3/8	6 3/8	6 3/8
Displacement—Cubic Inches	189.8	189.8	189.8	302.9	302.9	302.9	360.5	360.5	360.5
Compression Ratio	7.0:1	4.85:1	8.75:1	6.2:1	4.6:1	8.1:1	6.11:1	4.6:1	8.0:1
Compression Pressure @ Cranking Speed	145	84	200	132	84	185	129	86	194
Pistons Removed From	Front	Front	Front	Front	Front	Front	Front	Front	Front
Main and Rod Bearings Adjustable.	No	No	No	No	No	No	No	No	No
Cylinders Sleeved	No	No	No	No	No	No	No	No	No
Forward Speeds	6	6	6	6	6	6	6	6	6
Number of Main Bearings	2	2	2	2	2	2	2	2	2
Generator and Starter Make	D-R	D-R	D-R	D-R	D-R	D-R	D-R	D-R	D-R

TUNE-UP

	Gasoline	All-Fuel	LP-Gas	Gasoline	All-Fuel	LP-Gas	Gasoline	All-Fuel	LP-Gas
Valve Tappet Gap (Hot)	0.020	0.020	0.020	0.020	0.020	0.020	0.020	0.020	0.020
Ignition Distributor Make	D-R	D-R	D-R	D-R	D-R	D-R	D-R	D-R	D-R
Ignition Distributor Model	1112569	1112569	1112569	1112576	1112576	1112576	1112576	1112576	1112576
Breaker Contact Gap	0.022	0.022	0.022	0.022	0.022	0.022	0.022	0.022	0.022
Ignition Timing—Retard	TDC	TDC	TDC	TDC	TDC	TDC	TDC	TDC	TDC
Ignition Timing—Advanced	20°B	20°B	20°B	20°B	20°B	20°B	20°B	20°B	20°B
Flywheel Mark Indicating:									
Retard Timing	TDC	TDC	TDC	TDC	TDC	TDC	TDC	TDC	TDC
Advanced Timing	20	20	20	20	20	20	20	20	20
Spark Plug Size	18 mm	18 mm	18 mm	18 mm	18 mm	18 mm	18 mm	18 mm	18 mm
Spark Plug Electrode Gap	0.030	0.030	0.030	0.030	0.030	0.030	0.030	0.030	0.030
Carburetor Make	M-S	M-S	Own	M-S	M-S	Own	M-S	M-S	Own
Carburetor Float Setting	——— ¾-inch, measured from bowl gasket seat in casting to bottom of float———								
Engine Rated RPM	1325	1325	1325	1125	1125	1125	1125	1125	1125
Engine High Idle RPM	1460	1460	1460	1260	1260	1260	1260	1260	1260
Engine Slow Idle RPM	600	600	600	600	600	600	600	600	600

SIZES—CAPACITIES—CLEARANCES

(Clearances in Thousandths)

	Gasoline	All-Fuel	LP-Gas	Gasoline	All-Fuel	LP-Gas	Gasoline	All-Fuel	LP-Gas
Crankshaft Journal Diameter (Right)	2.874	2.874	2.874	3.9995	3.9995	3.9995	3.9995	3.9995	3.9995
Crankshaft Journal Diameter (Left)	2.249	2.249	2.249	2.7495	2.7495	2.7495	3.2495	3.2495	3.2495
Pulley Journal Diameter	2.182	2.182	2.182	2.433	2.433	2.433	2.433	2.433	2.433
Crankpin Diameter	2.8745	2.8745	2.8745	3.3743	3.3743	3.3743	3.7485	3.7485	3.7485
Piston Pin Diameter	1.4168	1.4168	1.4168	1.74975	1.74975	1.74975	2.35475	2.35475	2.35475
Valve Stem Diameter	0.372	0.372	0.372	0.4965	0.4965	0.4965	0.4965	0.4965	0.4965
Main Bearing Running Clearance:									
(Right)	4-6	4-6	4-6	5-7	5-7	5-7	5-7	5-7	5-7
(Left)	4-6	4-6	4-6	4-6	4-6	4-6	4.5-6.5	4.5-6.5	4.5-6.5
Pulley Bushing Running Clearance	3.5-4.5	3.5-4.5	3.5-4.5	3.5-5.5	3.5-5.5	3.5-5.5	3.5-5.5	3.5-5.5	3.5-5.5
Rod Bearing Running Clearance	1-4	1-4	1-4	1.1-4.1	1.1-4.1	1.1-4.1	2-5	2-5	2-5
Piston Skirt Clearance	——————Refer to paragraph 51——————								
Crankshaft End Play	5-10	5-10	5-10	5-10	5-10	5-10	5-10	5-10	5-10
Cooling System—Gallons	4 1/2*	4 1/2*	4 1/2*	6 1/2	6 1/2	6 1/2	7 1/8	7 1/8	7 1/8
Crankcase Oil—Quarts	7	7	7	8	8	8	10	10	10
Transmission and Differential—Qts.	16	16	16	24	24	24	32	32	32
Powershaft Clutch—Quarts	4	4	4	4	4	4	4 1/2	4 1/2	4 1/2
"Powr-Trol"—Quarts	10	10	10	11	11	11	13	13	13
Remote Cylinders, Each—Quarts	1	1	1	1	1	1	1	1	1
First Reduction Gear Cover—Quarts	1	1	1	1 1/2	1 1/2	1 1/2	1 1/2	1 1/2	1 1/2
Power Steering Reservoir—Quarts	5	5	5	5	5	5	5	5	5

TIGHTENING TORQUES—FT.-LBS

	Gasoline	All-Fuel	LP-Gas	Gasoline	All-Fuel	LP-Gas	Gasoline	All-Fuel	LP-Gas
Cylinder Head	104	104	104	150	150	150	275	275	275
Connecting Rod Nuts	85	85	85	105	105	105	85	85	85
Main Bearing Cap Screws	100	100	100	150	150	150	150	150	150
Spark Plugs	35	35	35	35	35	35	35	35	35
Cylinder Block to Main Case	166	166	166	175	175	175	275	275	275
Flywheel Bolts	150	150	150	275	275	275	275	275	275
Brake Pedal Shaft Nut	63	63	63	63	63	63	63	63	63
Rear Axle Housing to Main Case Cap Screws	112	112	112	167	167	167	275	275	275
Powershaft Clutch Housing	56	56	56	56	56	56	56	56	56
Rear Axle Housing to Main Case Nuts	....	....	....	....	....	....	150	150	150
Brake Lever Shaft Nut	63	63	63	63	63	63	63	63	63

*4 gals. for series 530.

FRONT SYSTEM—Tricycle Type

Tricycle type tractors, are available with a convertible type, two-piece pedestal as shown in Fig. JD1750. The available convertible front systems include a fork mounted single front wheel, dual wheels of the conventional knuckle mounted type or dual wheels of the "Roll-O-Matic" type. Refer to Fig. JD1751.

STEERING SPINDLE AND PEDESTAL
(Manual Steering)

1. R&R AND OVERHAUL. To remove the steering spindle (16—Fig. JD1750), first remove grille and on series 520, 620 and 720, remove steering wheel and the steering wheel Wood-

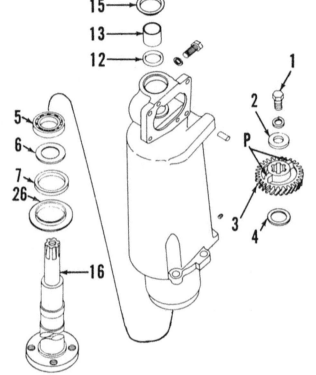

Fig. JD1750 — Exploded view of convertible two-piece pedestal. Spindle end play is controlled by washers (4).

1. Cap screw
2. Washer
3. Steering gear
4. Adjusting washers
5. Lower bearing
6. Washer
7. Cork washer
12. "O" ring
13. Bushing
15. Cup plug
16. Spindle
26. Retainer

ruff key. On all models, unbolt baffle plate from the radiator top tank. Remove cap screws retaining the steering worm housing to the pedestal and bump housings apart. Withdraw the steering worm and housing assembly from tractor. Remove cup plug from top of pedestal and remove cap screw retaining steering gear to spindle. Raise front of tractor and unbolt the lower pedestal extension and wheels assembly from the steering spindle. Using a knocker tool or brass drift, bump spindle down and out steering gear. Withdraw gear and save the adjusting washers which are located under the gear.

The spindle lower bearing (5) can be removed by using a suitable puller after removing retainer (26).

2. Inside diameter of a new spindle upper bushing (13) is 1.4995-1.5005 and the diameter of the steering spindle upper bearing surface should be 1.497-1.498. When installing the bushing, the beveled edge should be toward bottom of pedestal, the split should be toward the steering worm and top of bushing should be flush with top of bushing bore. Check the installed bushing to make certain that the steering spindle has the recommended clearance of 0.0015-0.0035 in the bushing. If there is evidence of oil

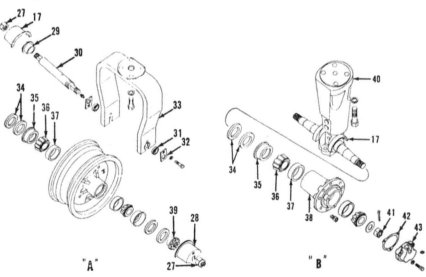

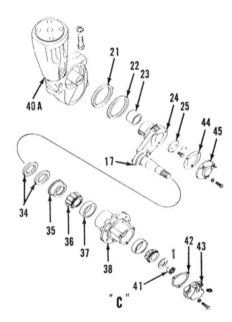

Fig. JD1751—Tricycle type front systems used with the pedestal shown in Fig. JD1750. A. Single front wheel. B. Conventional knuckle mounted dual front wheels. C. "Roll-O-Matic."

17. Dust excluder	25. Thrust washer	33. Yoke	41. Nut
21. Felt washer	27. Nut	34. Felt washers	42. Gasket
22. Retainer	29. Bearing spacer	35. Retainer	43. Cap
23. Bushing	30. Axle	36. Bearing cone	44. Gasket
24. "Roll-O-Matic"	31. Washer	37. Bearing cup	45. Cap
knuckle	32. Axle lock plate	38. Hub	
		40. Pedestal extension	
		(except "Roll-	
		O-Matic")	
		40A. Pedestal	
		extension	
		("Roll-O-Matic")	

leakage to the lower part of the pedestal, renew "O" ring (12) which is located just below the bushing (13).

3. When installing the steering spindle, observe the following:

(a) Be careful not to damage the "O" ring (12).

(b) Vary the number of spacer washers (4) to give the spindle an end play of 0.004-0.021.

Fig. JD1752 — "Roll-O-Matic" knuckle, showing the proper installation of bushings. Notice that open end of oil grooves in bushings is toward the 1/32-1/16-inch gap between the bushings.

(c) With front wheel (or wheels) pointing straight ahead, the hub projections (P) on steering gear should be parallel to center line of tractor.

Steering spindle to wheel extension bolts should be tightened to a torque of 275 ft.-lbs.

"ROLL-O-MATIC"

4. OVERHAUL. The "Roll-O-Matic" unit can be overhauled without removing the unit from the tractor.

Support front of tractor and remove wheel and hub units. Remove knuckle caps (45—Fig. JD1751C). Unbolt and remove thrust washers (25). Pull knuckle and gear units from housing and remove felt washer (21). On regular duty "Roll-O-Matic" units, check

Fig. JD1753—When assembling the "Roll-O-Matic" unit make certain that gears are meshed so that timing marks (M) are in register.

the removed parts against the values which follow:

Knuckle Bushing Inside Diameter
Series 520-530 . . . . .1.623-1.625
Series 620-630-720-730 . .1.873-1.875

Knuckle Shaft Outside Diameter
Series 520-530 . . . . .1.620-1.622
Series 620-630-720-730 . .1.870-1.872

Thickness of Thrust Washers (25)
All Series .0.156

On heavy duty "Roll-O-Matic" units, check the removed parts against the values which follow:

Knuckle Bushing Inside Diameter
Series 520-530 . . . . .1.873-1.875
Series 620-630-720-730 . .2.127-2.129

Knuckle Shaft Outside Diameter
Series 520-530 . . . . .1.870-1.872
Series 620-630-720-730 . .2.124-2.126

Thickness of Thrust Washers (25)
Series 520-530 . . . . .0.156
Series 620-630-720-730 . . . . . .0.187

Install knuckle bushings (23) with open end of oil groove toward gap between bushings as shown in Fig. JD1752. When the bushings are properly installed, there should be a gap of 1/32-1/16 inch between the bushings to allow grease from the fittings to enter grooves in bushings.

Soak felt washers (21—Fig. JD-1751C) in engine oil prior to installation. Install one of the knuckles so that wheel spindle extends behind the vertical steering spindle. Pack the "Roll-O-Matic" unit with wheel bearing grease and install the other knuckle so that timing marks on gears are in register as shown in Fig. JD1753.

FRONT SYSTEM—Axle Type

STEERING KNUCKLES

(Except Orchard and Standard)

5. R&R AND OVERHAUL. Before removing knuckle, raise front of tractor and check end play of knuckle post in the axle knee. If end play exceeds 0.036, renew thrust washers (12—Figs. JD1754 and 1755) when reassembling.

To remove either knuckle, remove wheel and hub assembly, disconnect drag link from steering arm and remove nut retaining steering arm to top of knuckle post. Using a knocker tool or brass drift, drive knuckle post free of steering arm. Inside diameter of new knuckle post bushings should be 1.504-1.506 and the knuckle post diameter should be 1.494-1.495.

When installing new bushings, press them in until flush with top and bottom of knee and make certain that oil groove in each bushing is in register

Fig. JD1754 — Exploded view of 48-80-inch adjustable axle knuckle, extension and associated linkage.

1. Knee extension
2. Steering arm
3. Knee
4. Washer
5. Knuckle bushings
6. Knuckle
7. Dust excluder
8. Dowel pin
9. Dust shield
12. Thrust washer
13. Drag link end
14. Extension bar
15. Outer drag link
16. Inner drag link
17. Center steering arm

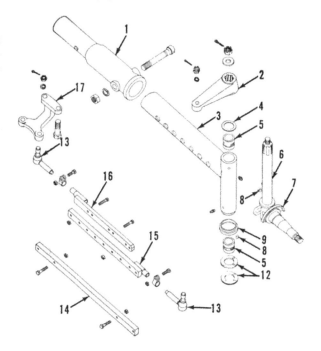

with fittings in knee. Ream the bushings after installation, if necessary, to give the knuckle post the recommended clearance of 0.009-0.012 in the bushings.

When reassembling, install dust shield (9) on knee. Install thrust washers (12) on knuckle post and install knuckle and post assembly into knee, making certain that dowel pin engages holes in thrust washers.

Series 620-630-720-730 (Orchard and Standard)

5A. **R&R AND OVERHAUL.** To remove the steering knuckles (8—Figs. JD1756, 1757 or 1758), remove the front wheel and hub units, disconnect the steering arms from knuckles and remove knuckle caps (4). Remove the taper bolt retaining nuts and drive the taper bolts forward and out of axle. Using a drift, bump spindle pins (5) out of axle and knuckle.

Spindle bushings (7) are pre-sized and if not distorted during installation, will require no final sizing. Recommended clearance between new spindle pins and knuckle bushings is 0.002-0.005.

DRAG LINKS AND TOE-IN

6. An adjustable ball socket is fitted to both ends of each drag link on some models and the link ends should be adjusted so they have no end play, yet do not bind. On other models, the tie rod and drag link ends are of the non-adjustable automotive type.

With front wheels pointing straight ahead, toe-in should be $\frac{1}{8}$-$\frac{3}{16}$ inch. If the adjustment is not as specified, adjust the length of each drag link or tie-rod an equal amount, until proper adjustment is obtained.

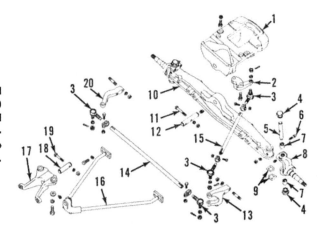

Fig. JD1756 — Exploded view of 620, 630, 720 and 730 standard fixed tread front axle and associated linkage. Refer to legend under Fig. JD1757.

AXLE PIVOT PINS AND BUSHINGS (Except Orchard and Standard)

7. To renew the axle pivot pins and bushings, support front of tractor and disconnect the center steering arm from the steering spindle. Unbolt axle pivot bracket from front end support and roll axle, pivot bracket and wheels assembly forward and away from tractor. Remove the retaining bolt and withdraw pivot bracket from axle. Inside diameter of new pivot pin bushings is 1.504-1.506 and diameter of new pivot pins is 1.494-1.495.

When installing a new bushing in axle or pivot bracket, press or drive bushing in until bushings are flush with castings and oil groove in each bushing is in register with grease fittings in axle and pivot bracket. Ream the bushings after installation, if necessary, to provide a clearance of 0.009-0.012 for the pivot pins. Pivot pins can be driven from axle and pivot bracket and new ones can be driven or pressed in. Be sure to measure pivot pin ex-

tension from axle or pivot bracket and install new ones to the same dimension.

Reassemble and reinstall the axle and pivot bracket assembly by reversing the removal procedure.

Series 620-630-720-730 (Orchard and Standard)

7A. Axle and radius rod are unbushed at pivot pin locations. The procedure for renewing the pins is evident.

STEERING SPINDLE AND PEDESTAL (Manual Steering)

8. **R&R AND OVERHAUL.** To remove the steering spindle (16—Fig. JD1750), first remove grille and on series 520, 620 and 720, remove steering wheel and the steering wheel Woodruff key. On all models, unbolt baffle plate from radiator top tank. Remove cap screws retaining the steering worm housing to the pedestal and bump the housings apart. Withdraw the steering worm and housing assembly from tractor. Support front of tractor and disconnect center steering arm from the steering spindle. Unbolt axle pivot bracket from front end support and roll axle, pivot bracket and wheels assembly forward and away from tractor. Remove cup plug from top of pedestal and remove cap screw retaining steering gear to spindle. Using a knocker tool or brass drift, bump spindle down and out of steering gear. Withdraw gear and save the adjusting washers which are located under the gear.

The spindle lower bearing (5) can be removed by using a suitable puller after removing retainer (26).

Inside diameter of a new spindle upper bushing (13) is 1.4995-1.5005 and the diameter of the steering spindle upper bearing surface should be 1.497-1.498. When installing the bushing, the beveled edge should be toward bottom of pedestal, the split should

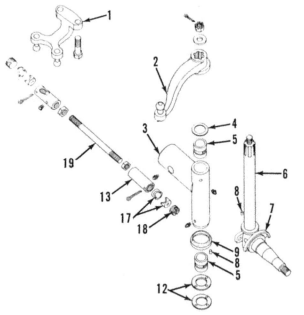

Fig. JD1755 — Exploded view of 38-inch fixed tread axle knuckle, knee and associated linkage. The Hi-Crop models are similarly constructed.

2. Steering arm
3. Knee
4. Washer
5. Knuckle bushings
6. Knuckle
7. Dust excluder
8. Dowel pin
9. Dust shield
12. Thrust washer
13. Drag link end
17. Stud bearings
18. Screw plug
19. Drag link rod

be toward the steering worm and top of bushing should be flush with top of bushing bore. Check the installed bushing to make certain that the steering spindle has the recommended clearance of 0.0015-0.0035 in the bushing. If there is evidence of oil leakage to the lower part of the pedestal, renew "O" ring (12) which is located just below the bushing (13).

When installing the steering spindle, observe the following:

(a) Be careful not to damage the "O" ring (12) when installing the steering spindle.

(b) Vary the number of spacer washers (4) to give the spindle an end play of 0.004-0.021.

(c) With steering spindle positioned so that when the center steering arm is connected and the front wheels are pointing straight ahead, the hub projections (P) on steering gear should be parallel to center line of tractor.

Steering spindle to center steering arm screws should be tightened to a torque of 275 ft.-lbs.

MANUAL STEERING SYSTEM

(Except Orchard)

For the purposes of this discussion, the steering mechanism will include the steering worm and shaft and the steering gear. For R&R and overhaul of the steering spindle, refer to paragraphs 1 or 8.

9. **ADJUSTMENT.** Three adjustments are provided on the steering mechanism: (1) steering spindle shaft end play; (2) steering (worm) shaft end play; (3) backlash between the worm and steering gear.

10. **STEERING SPINDLE SHAFT END PLAY.** The desired steering spindle shaft end play of 0.004-0.021 is controlled by adjusting washers (4—Fig. JD1761) which are located under the steering gear (3). To make the adjustment, it is necessary to remove the steering gear from tractor as outlined in paragraph 14.

11. **STEERING (WORM) SHAFT END PLAY.** The desired steering (worm) shaft end play of 0.001-0.004 is controlled by shims (81—Figs. JD1760

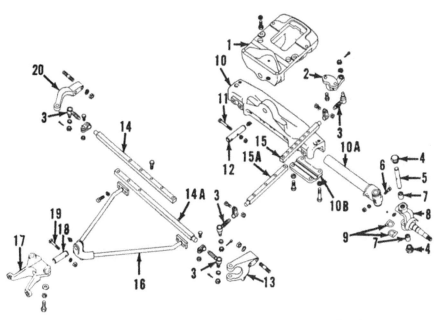

Fig. JD1757—Exploded view of 620, 630, 720 and 730 standard adjustable tread front axle and associated linkage.

1. Axle pivot bracket	9. Thrust washers	14A. Tie rod extension
2. Center steering arm	10. Front axle	15. Drag link
3. Tie rod and	10A. Axle extension	15A. Drag link extension
drag link ends	10B. Extension clamp	16. Radius rod
4. Knuckle cap	11. Bolt	17. Radius rod pivot bracket
5. Spindle pin	12. Axle pivot pin	18. Radius rod pivot pin
6. Taper bolt	13. Right steering arm	19. Bolt
7. Spindle bushings	14. Tie rod	20. Left steering arm
8. Steering knuckle		

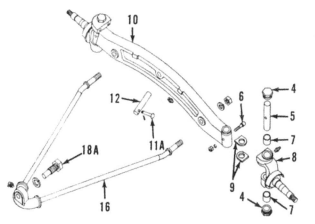

Fig. JD1758 — Exploded view of front axle, knuckle and associated linkage used on 620 orchard tractors.

4. Knuckle caps
5. Spindle pin
6. Taper bolt
7. Spindle bushings
8. Steering knuckle
9. Thrust washers
10. Front axle
11A. Rivet
12. Axle pivot pin
16. Radius rod
18A. Radius rod pivot bolt

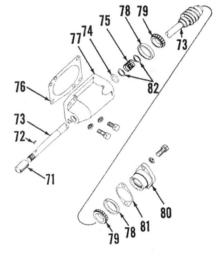

Fig. JD1760 — 520, 620 and 720 manual steering worm shaft and associated parts.

71. Bushing	77. Worm housing
72. Woodruff key	78. Bearing cup
73. Worm shaft	79. Bearing cone
74. "O" ring packing	80. Bearing housing
75. Spring	81. Shims
76. Shims	82. Washer

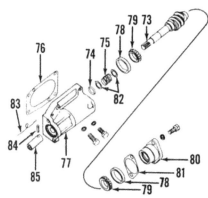

Fig. JD1760A — 530, 630 and 730 manual steering worm shaft and associated parts.

73. Worm shaft	80. Bearing housing
74. "O" ring packing	81. Shims
75. Spring	82. Washer
76. Shims	83. Wire
77. Worm housing	84. Roll pin
78. Bearing cup	85. Coupling
79. Bearing cone	

or JD1760A) which are located under the worm shaft front bearing housing. Attach dial indicator to steering worm shaft in a suitable manner, move worm shaft back and forth and check the end play. If end play is not as specified, remove the worm shaft front bearing housing and add or remove shims until proper adjustment is obtained.

12. BACKLASH. The desired backlash between the steering gear and worm is ½-1 inch when measured at rim of steering wheel, and is controlled by shims (76—Figs. JD1760 or JD1760A) which are located between pedestal and worm housing. If backlash is not as specified, remove grille and unbolt baffle plate from the radiator top tank. Unbolt worm housing from pedestal, bump housings apart and add or remove shims (76) until backlash is as specified. Shims are available in thicknesses of 0.005 & 0.015. If excessive backlash cannot be eliminated, it will be necessary to renew the worm and steering gear; or, reposition the steering gear on the steering spindle so as to bring unworn teeth into mesh. Refer to paragraph 14. Note: On series 530, 630 and 730, be careful not to confuse wear in the steering shaft U-joints or couplings with gear unit backlash.

13. OVERHAUL. The steering mechanism can be overhauled without removing pedestal from tractor, as follows:

14. STEERING GEAR. To remove steering gear (3—Fig. JD1761), first remove grille and on series 520, 620 and

720, remove steering wheel and steering wheel Woodruff key. On all models, unbolt baffle plate from radiator top tank and remove cup plug from top of pedestal. Unbolt worm housing from pedestal and bump the housings apart. Withdraw wormshaft and housing assembly from tractor. On axle type tractors, disconnect center steering arm from steering spindle and unbolt axle pivot bracket from front end support. On all models, raise front of tractor, remove cap screw retaining steering gear to spindle and using a knocker tool or brass drift, bump spindle down and out of steering gear. Reinstall the steering gear by reversing the removal procedure and observe the following:

(a) If the same gear is being reinstalled, position the gear so that unworn teeth will mesh with the worm.

(b) Vary the number of spacer washers (4) to give the spindle an end play of 0.004-0.021.

(c) With front wheel (or wheels) pointing straight ahead, the hub projections (P) on steering gear should be parallel to center line of tractor.

15. STEERING WORM. Overhaul of the steering (worm) shaft and components is accomplished as follows: Remove grille and on series 520, 620 and 720, remove steering wheel and the steering wheel Woodruff key. Remove the worm shaft front bearing housing (80—Figs. JD1760 or 1760A) and turn

worm shaft forward and out of housing. The worm shaft housing can be removed from pedestal at this time. The need and procedure for further disassembly is evident after an examination of the unit. Inspect "O" ring seal (74) and renew if damaged.

When reassembling, adjust the worm shaft end play and the gear unit backlash as outlined in paragraphs 11 and 12.

Series 620 (Orchard)

15A. ADJUSTMENT. Three adjustments can be made on the steering gear unit: (1) the steering worm shaft end play; (2) the vertical shaft end play; and (3) the backlash between the steering worm and the steering gear. Before adjusting the steering gear unit, it is recommended that the front end of the tractor be raised so that all unnecessary load is removed from the steering mechanism.

15B. STEERING WORM SHAFT END PLAY. To adjust the steering worm shaft end play, loosen clamp bolt (13—Fig. JD1762) and turn adjusting plug (3) either way as required until the worm shaft has zero end play, yet turns freely.

15C. VERTICAL STEERING SHAFT END PLAY. Loosen jam nut (20—Fig. JD1763) and turn adjusting screw (19) down until it is tight against top of vertical steering shaft. Then back off the adjusting screw ⅛ turn and secure it by tightening jam nut (20).

15D. BACKLASH ADJUSTMENT. Remove cap screws retaining the steering worm shaft housing (11—Fig. JD1762) to the steering gear housing and separate the two housings. Add or

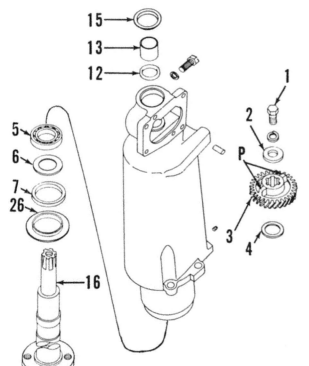

Fig. JD1761 — Exploded view of convertible two-piece pedestal. Spindle end play is controlled by washers (4).

1. Cap screw
2. Washer
3. Steering gear
4. Adjusting washers
5. Lower bearing
6. Washer
7. Cork washer
12. "O" ring
13. Bushing
15. Cup plug
16. Spindle
26. Retainer

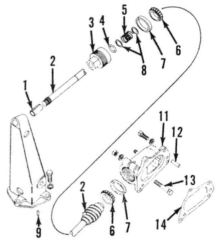

Fig. JD1762 — Exploded view of 620 orchard steering worm shaft, housing and associated parts.

1. Bushing	8. Washers
3. Adjusting plug	9. Dowel pin
4. "O" ring	11. Worm housing
5. Spring	12. Dowel pin
6. Bearing cone	13. Clamp bolt
7 Bearing cup	14. Shims

remove shims (14), which are located between the two housings, until there is ½-1 inch free movement measured on the rim of the steering wheel. Note: Check to make certain that the housing locating dowels are in place before tightening the cap screws.

Check for binding through entire range of steering wheel travel. If the steering gear binds, or has an excessive amount of backlash in any position, it will be necessary to renew the worm and/or gear; or, re-position them so as to bring unworn teeth into mesh.

15E. OVERHAUL. To overhaul the steering gear unit, it is first necessary to remove the unit from tractor. Steering gear shaft bushing (24—Fig. JD1763) which is driven into the bottom portion of the tractor main case, can be renewed at this time. Ream the bushing after installation, if necessary, to provide 0.003-0.006 clearance for the vertical steering shaft.

Procedure for disassembly and overhaul of the unit is obvious.

15F. STEERING SHAFT TUBE — RENEW. To renew the vertical steering shaft tube (22—Fig. JD1763), first remove the steering gear unit from tractor. Drive bushing (24) out through bottom of main case and drive the tube out through top of main case. Install new tube from above and using a piloted driver, drive the tube down

until lower shoulder of tube bottoms in the main case. Install new bushing (24) and ream the bushing after installation, if necessary, to provide 0.003-0.006 clearance for the vertical steering shaft.

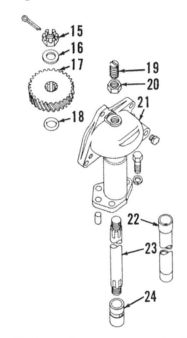

Fig. JD1763 — Steering gear and housing units used on 620 orchard tractors.

16. Washer	21. Gear housing
17. Steering gear	22. Steering shaft
18. "O" ring packing	tube
19. Set screw	23. Steering shaft
20. Jam nut	24. Bushing

POWER STEERING SYSTEM

(Except Orchard)

Note: The maintenance of absolute cleanliness of all parts is of utmost importance in the operation and servicing of the hydraulic power steering system. Of equal importance is the avoidance of nicks or burrs on any of the working parts.

LUBRICATION

16. It is recommended that the power steering system be drained (but not flushed) once a year or every 1,000 hours. To drain the system, turn the steering wheel to the extreme left position and remove the grille. Remove the reservoir drain plug and allow system to drain. To remove the additional oil remaining in the steering cylinder, disconnect the front oil line from the control valve, pivot the line forward and turn the steering wheel to the extreme right position. Refill the system with five U.S. quarts of John Deere power steering oil.

TROUBLE-SHOOTING

17. The following paragraphs outline the possible causes and remedies for troubles in the power steering system.

17A. DRIFTING TO EITHER SIDE could be caused by:

1. Valve housing not in correct relation to worm shaft housing. Center the valve housing as in paragraph 23.

17B. HARD STEERING could be caused by:

1. Insufficient volume of oil flowing to steering valve from flow control valve. Adjust the flow control valve as in paragraph 22.
2. Excessive tension on the centering cam spring. Adjust the spring as in paragraph 25C or 26C.
3. Insufficient end play or binding in the worm shaft bearings. See paragraph 25 or 26.

4. Insufficient backlash between the steering worm and gear. See paragraph 25B or 26B.
5. Excessive leakage past the steering vane seals in cylinder. To check, remove grille, disconnect the front oil line from steering valve, pivot the oil line forward and tighten the lower oil line connection. With the engine running at fast idle, hold the steering wheel to the extreme right turn position and measure the leakage from the oil line as shown in Fig. JD1764. Oil leakage should not exceed 1 quart in ½ minute. If leakage is excessive, renew the vane seals. Refer to paragraph 29B.
6. Binding in the rear steering shaft support bushing.
7. Pedestal improperly aligned. See paragraph 25A or 26A.
8. Insufficient pump pressure. Refer to paragraph 19.
9. Insufficient oil in system.
10. Foaming oil in system. Refer to paragraph 17D.

17C. EXCESSSIVE INSTABILITY OF FRONT WHEELS. This condition is often referred to as shimmy or flutter. In some cases, flutter cannot be entirely eliminated, but the unit can be adjusted to the point where it is not objectionable. Possible causes of instability are:

1. Excessive volume of oil flowing from flow control valve to steering valve. Adjust the flow control valve as in paragraph 22.

Fig. JD1764 — Checking for leakage past the power steering cylinder vane.

2. Actuating sleeve set screw too loose in helix slot. Refer to paragraph 25.

3. Worn point on actuating sleeve set screw and/or damaged helix slot in steering worm shaft.

4. Insufficient tension on cam spring. Refer to paragraph 25C or 26C.

5. Excessive end play in the steering worm shaft, paragraph 25 or 26.

6. Excessive backlash between the steering worm and gear. Refer to paragraph 25B or 26B.

7. Unbalanced front wheels.

8. Loose or worn front wheel bearings.

9. Loose or worn "Roll-O-Matic" assembly.

10. Use of front wheel weights rather than front end weights.

17D. OVERFLOW OR FOAMING OF OIL could be caused by:

1. Air leak in system. To check, apply a light coat of oil to sealing surfaces and observe for leaks.

2. Wrong type oil. Use only John Deere power steering oil.

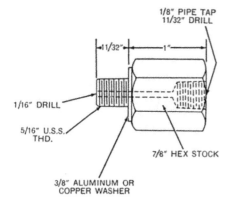

Fig. JD1764A—Adapter which can be made to accommodate a pressure gage in the steering valve housing.

17E. LOCKING could be caused by:

1. Scored worm, or worm bearings adjusted too tight. Refer to paragraph 25 or 26.

2. Steering valve arm interference in groove of actuating sleeve.

3. Steering gear loose on vertical spindle. Tighten the cap screw to a torque of 190 ft.-lbs.

4. Bent steering spindle or scored steering spindle bearings. Refer to paragraph 29B.

5. Loose or broken movable vane retaining cap screws. Screws should be tightened to a torque of 208 ft.-lbs. Refer to paragraph 29B.

6. Insufficient clearance between actuating sleeve and the steering worm shaft or cam. Refer to paragraph 25C or 26C.

17F. VARIATION IN STEERING EFFORT WHEN TURNING IN ONE DIRECTION could be caused by a bent steering spindle. Refer to paragraph 29B.

SYSTEM OPERATING PRESSURE

19. The system relief valve is mounted in the steering valve housing as shown in Fig. JD1769. To check the system operating pressure, remove grille and using the adapter (Fig. JD-1764A) install a suitable pressure gage (at least 1500 psi capacity) in flow control stop screw hole as shown in Fig. JD1764B. With the engine running at fast idle rpm, cramp the front wheels to the extreme right or extreme left position, and observe the highest pressure reading which should be 1170-1210 psi.

If the operating pressure is not as specified, vary the number of washers (19—Fig. JD1769). If the addition of washers will not increase the pressure to within the specified limits, look for a faulty pump.

PUMP

20. REMOVE AND REINSTALL. To remove the power steering pump, first remove the grille and proceed as follows: On series 520, 620 and 720, disconnect the steering shaft coupling and remove the steering shaft. On all models, remove the hood. Drain the power steering reservoir and remove pump pressure and suction oil lines. Loosen and disconnect the fan belt. Unbolt the fan shaft tube flange from pump and pump from mounting bracket.

Move the pump and fan assembly forward until the fan shaft coupling is just exposed. While holding the coupling rearward, move the pump assembly forward until the pump shaft is disengaged from the coupling.

Note: It coupling is not held rearward while pump is moved forward, the fan shaft may come out of the rear coupling. If the shaft comes out of the rear coupling, it is difficult to replace.

Move rear of pump toward right side of tractor, tip top of pump toward right side and withdraw the pump and fan assembly from left side of tractor.

Note: In some individual cases it is more convenient and often time will be saved by removing the fan blades before unbolting the pump from its mounting bracket.

Install the pump by reversing the removal procedure and fill the pump with oil before connecting the oil lines.

21. OVERHAUL. With the pump removed from tractor, thoroughly clean the exterior surfaces to remove any accumulation of dirt or other foreign material. Remove the pump housing retaining cap screws and using a plastic hammer, bump the pump housing

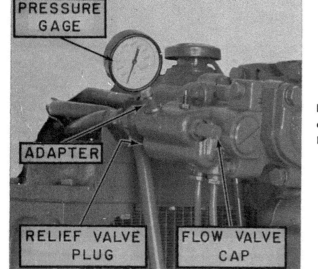

Fig. JD1764B — Adapter and pressure gage installation for checking the power steering system.

Fig. JD1765 — Compressing the fan shaft spring (4) to permit the removal of locks (9) and keeper (8).

from its locating dowels. Remove the pump body, follower gear and drive gear. Refer to Fig. JD1767. Extract the drive gear Woodruff key (2) from the fan shaft.

Place the pump cover and fan assembly on a press, depress the fan against spring pressure as shown in Fig. JD1765 and remove the split cone locks (9) and keeper (8). Remove as-

sembly from press and withdraw the fan, pulley, spring and drive parts from the shaft. Press the pump drive shaft rearward out of pump cover, extract the bearing retaining snap ring and remove bearing from the cover. Inspect all parts and renew any which are questionable. Bushings in pump body and cover should be pressed in with a piloted arbor. Oil seal in pump

cover should be pressed in not more than $\frac{1}{16}$-inch below bottom of bearing bore. Pump housing oil seal (24) is replaced by two seals and a spacer plate that are installed as follows: Install inner seal with lip facing inward (toward pump), followed by the spacer, then the outer seal with lip facing outward (toward engine).

To reassemble the pump, install the cover bearing and snap ring. Press the pump shaft into bearing from gear side of cover until shoulder on shaft seats against the bearing race. Install the two assembly cap screws in cover, then install the driving pulley, spring and fan parts in the sequence shown in Fig. JD1766. Install Woodruff key in the drive shaft and slide the drive gear into position. Drive gear must float on shaft and not bind on the Woodruff key. Install the follower gear, coat the mating surfaces of the pump cover and body with shellac and install the body so that edge of body are concentric with edges of cover. Coat the mating surfaces of the pump body and housing with shellac, install the housing and tighten the assembly cap screws securely.

STEERING VALVES

22. **ADJUST FLOW CONTROL VALVE.** Turning the flow control valve adjusting screw inward will cause faster, easier steering action; but, the increase in steering speed can result in a decrease of front wheel stability (increased tendency of front wheels to flutter).

To make the adjustment, operate the steering system until the oil is at normal operating temperature; then turn the adjusting screw in until the fastest turning speed is obtained without causing an objectionable amount of front wheel flutter. Refer to Fig. JD1768.

Note: Front wheel instability (flutter) can also be caused by improper adjustment of the steering worm and sleeve assembly. Refer to paragraph 25 or 26.

Fig. JD1766 — Exploded view of power steering pump used on all except orchard models. System relief valve is located within the valve housing. Seal (24) is replaced by two seals and a spacer.

1. Fan shaft
2. Woodruff key
3. Pulley
4. Spring
5. Friction disc
6. Friction washer
7. Drive cup
8. Fan keeper
9. Locks
10. Snap ring
11. Bearing
12. Oil seal
13. Dowel pin
14. Pump body
15. Pumping gears
16. Idler gear shaft
17. Bushing
24. Oil seal
25. Pump housing
26. Bushing
27. Woodruff key
28. Bushing
29. Pump cover

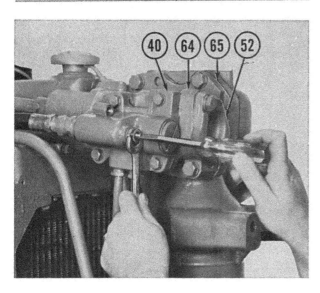

Fig. JD1767 — Partially disassembled view of power steering pump. Be sure to remove Woodruff key (2) before attempting to remove fan shaft (1).

Fig. JD1768 — Adjusting the power steering flow control valve. To make the adjustment, turn the screw in to obtain the fastest turning speed without causing an objectionable amount of front wheel flutter.

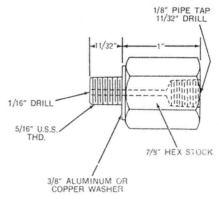

1/8" PIPE TAP
11/32" DRILL

1/16" DRILL

5/16" U.S.S. THD.

7/8" HEX STOCK

3/8" ALUMINUM OR COPPER WASHER

Fig. JD1768A—Adapter which can be made to accommodate a pressure gage in the steering valve housing.

23. CENTERING VALVE HOUSING.
Housing (40—Figs. JD1768 and 1769) is properly centered when the effort required to turn the steering wheel in one direction is exactly the same as the effort required to turn the steering wheel in the other direction. The adjustment is made by loosening the valve housing retaining cap screws, tapping the housing either way as required and then tightening the cap screws.

If the wheel turns harder to the left, tap the housing rearward. Conversely, if the wheel turns harder to the right, tap the valve housing forward.

To accurately center the valve housing using a pressure gage, use the adapter (Fig. JD1768A) and install a suitable pressure gage (at least 1500 psi) in flow control stop screw hole as shown in Fig. JD1768B. With the engine running at fast idle speed and

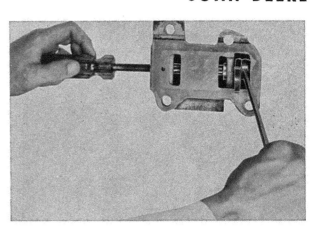

Fig. JD1770 — Removing bolt retaining the steering valve arm to the steering valve.

front wheels in the straight ahead position, loosen the steering valve attaching screws and tap the housing to the front or rear until the lowest pressure reading is obtained; then tighten the housing retaining cap screws.

24. REMOVE AND REINSTALL.
To remove the valve housing, remove

grille, drain the power steering reservoir and disconnect the oil lines from valve housing. To facilitate reinstallation, mark the relative position of the valve housing with respect to the worm housing with a scribed line. Unbolt and remove the valve housing assembly from tractor.

24A. To install the valve housing, use a new gasket and tighten the assembly cap screws finger tight; also, be sure to align the previously affixed scribed lines. Reconnect the oil lines, fill the reservoir and start engine. Center the valve housing as in paragraph 23.

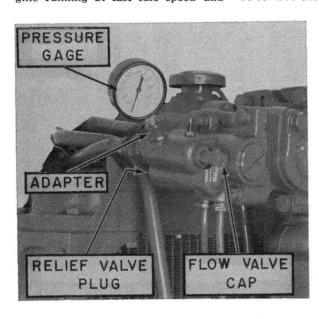

Fig. JD1768B — Adapter and pressure gage installation for checking the power steering system.

24B. OVERHAUL. With the valve housing removed, clean the unit and remove the flow control valve adjusting screw. Remove the union adapter from rear of housing and slide the flow control valve and spring out of the valve bore. Remove the end plug from housing and remove the steering valve arm by removing its retaining bolt as shown in Fig. JD1770. Remove the relief valve plug (21—Fig. JD1769) and withdraw the relief valve sleeve, valve and spring. Wash all parts in a suitable solvent and renew any that are nicked, grooved or worn.

Install the flow control valve spring, valve and adjusting screw, and initially adjust the valve as follows: Using a punch or similar tool, hold the valve against rear of flow valve passage and turn the adjusting screw in until it just contacts the spring; then turn the screw in two additional turns. Install the union adapter, flow valve and arm and end plug. Assemble the relief valve, using the same number of adjusting washers as were removed.

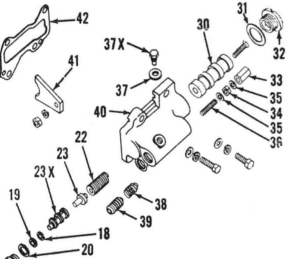

Fig. JD1769 — Exploded view of the steering valve housing.

18. Washers
19. Shim washers
20. Gasket
21. Plug
22. Spring
23. Relief valve
23X. Relief valve sleeve
30. Steering valve
31. Gasket
32. Plug
33. Cap nut
34. Jam nut
35. Washer
36. Adjusting screw
37. Copper washer
37X. Stop screw
38. Flow control valve spring
39. Flow control valve
41. Valve arm

With the valve housing installed, center the unit as in paragraph 23 and adjust the flow control valve as outlined in paragraph 22. Check the system operating pressure as outlined in paragraph 19.

STEERING WORM AND VALVE ACTUATING SLEEVE
Early Series 520-620-720

25. **END PLAY ADJUSTMENT.** Two end play adjustments are provided in the steering worm and valve actuating sleeve assembly. The valve actuating sleeve set screw must be adjusted to provide an end play of 0.001-0.002 between the cone point of the set screw and the helix slot in the steering worm shaft. Then, the number of shims between the worm housing and the front bearing housing should be varied to obtain a worm shaft end play of 0.001-0.004. The two adjustments will therefore provide a total end play for the worm and actuating sleeve assembly of 0.002-0.006.

The two adjustments can be approximated, but since the end play should be held as near as possible to the lower limit, it is recommended that a dial indicator be used as follows:

Remove grille and mount a dial indicator with contact button resting on rear end of the valve actuating sleeve as shown in Fig. JD1771. Remove the worm shaft front bearing housing and all of the adjusting shims. Reinstall the front bearing housing without the shims so that the worm shaft bearings have no end play. Remove the inspection hole plug and turn the steering wheel until the actuating sleeve set screw lines up with the inspection hole. Turn the set screw either way as required to obtain an end play of 0.001-0.002. It is desirable, however,

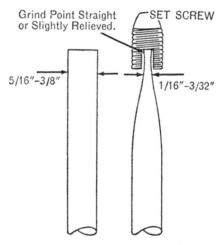

Fig. JD1772 — Dimensions for grinding a special screwdriver for adjusting the early production 520, 620 and 720 valve actuating sleeve set screw.

to hold the end play adjustment as close to 0.001 as possible.

Note: Adjustment of the set screw will be made easier by grinding a screw driver tip to the dimensions shown in Fig. JD1772. After adjustment, lock the screw in position by staking.

With the set screw adjusted, remove the worm shaft front bearing housing, install the previously removed shims and check the end play which should now be 0.002-0.006. Note: This 0.002-0.006 end play includes the 0.001-0.002 end play previously obtained by the actuating sleeve set screw adjustment. The number of shims under the front bearing housing should be varied to reduce the total end play as near to 0.002 as possible. Shims are available in various thicknesses.

25A. **STEERING SHAFT BIND.** To obtain proper steering action, there should be a minimum of binding in the steering shaft. To check for binding with engine stopped, turn the steering wheel in either direction without turning the front wheels. When the steering wheel is released,

spring pressure against the cam within the valve actuating sleeve should return the steering wheel to a neutral position.

If binding exists after the end play is adjusted as outlined in paragraph 25, check for misalignment of the steering shaft rear support. The support can be aligned after loosening the two retaining cap screws.

An improperly located pedestal can also cause binding. Beginning with the front center pedestal retaining cap screw, loosen every other cap screw and bump pedestal either way as shown in Fig. JD1773 until the steering shaft is free.

25B. **BACKLASH.** Backlash between the steering worm and gear is controlled by shims located between the worm shaft housing and the pedestal. To check the backlash, rotate the steering wheel back and forth without permitting the steering cams within the actuating sleeve to separate or the front wheels to turn. Backlash should be $\frac{3}{8}$-$\frac{3}{4}$ inch when measured at outer edge of steering wheel. Check the backlash with the front wheels in the extreme right, straight ahead and extreme left positions. Excessive variation in backlash between the three positions indicates a bent steering spindle.

25C. **R&R AND OVERHAUL.** To remove the steering worm and valve actuating sleeve, first remove the grille, disconnect the steering shaft coupling and drain the power steering oil reservoir. Disconnect the oil lines and remove the steering valve housing as outlined in paragraph 24. Unbolt and remove the worm shaft housing from pedestal, being careful not to damage or lose the backlash adjusting shims.

Remove oil seal housing (46—Fig. JD1774), actuating sleeve set screw (51) and actuating sleeve (48). When withdrawing the actuating sleeve, it

Fig. JD1771 — Adjusting the early production 520, 620 and 720 steering valve actuating sleeve set screw. Adjustment is best accomplished by grinding a screwdriver tip as shown in Fig. JD1772.

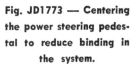

Fig. JD1773 — Centering the power steering pedestal to reduce binding in the system.

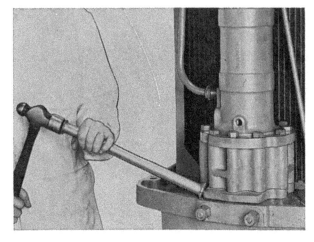

may be necessary to insert a punch through the sleeve hole as shown in Fig. JD1775 to position the cam rod spring so it will clear the internal pins in the sleeve. Unbolt the worm shaft front bearing housing, save the adjusting shims and withdraw the worm. Remove cam spring (61—Fig. JD1774) and cam (58) from rod.

Clean and inspect all parts and renew any that are damaged and cannot be reconditioned. Lip of oil seal (45) goes toward worm. Renew the worm shaft if it is scored or worn at cam end, helix slot or worm. Burrs can be removed from the worm shaft keyway or the surface over which the actuating sleeve operates by using a fine stone. Check the valve actuating sleeve pins for damage, and bore of sleeve for being scored. Check fit of sleeve on worm shaft. If sleeve is tight on shaft or if bore is scored, remove the dowel pins and hone the sleeve to insure a free fit. Renew the pins if they are worn or damaged. Renew the actuating sleeve set screw if it has a loose fit or if cone point is worn.

When reassembling, use a valve spring tester and measure the length of spring (61) at 70-80 lbs. pressure.

Note: If spring will not exert 70-80 lbs. pressure at 2⅞ inches or more for a 3-inch free length spring or 1⅞ inches or more for a 2⅛-inch free length spring, obtain a new spring and measure its length at 70-80 lbs. pressure.

Then install the spring on the cam rod and tighten the adjusting nut until spring is compressed to the same length which yielded 70-80 lbs.

Reassemble the worm and actuating sleeve and adjust the end play as in paragraph 25. Install the unit and ad-

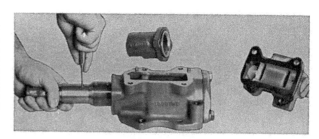

Fig. JD1775 — When removing the valve actuating sleeve on early production 520, 620 and 720 tractors, it is often necessary to use a punch and position the cam rod spring so it will clear the sleeve pins.

just the backlash as in paragraph 25B. Install the valve housing as outlined in paragraph 24A.

STEERING WORM AND VALVE ACTUATING SLEEVE
Late Series 520-620-720 and Series 530-630-730

26. END PLAY ADJUSTMENTS. Two end play adjustments are provided in the steering worm and valve actuating sleeve assembly. The number of shims between the worm housing and the front bearing housing should be varied to obtain a worm shaft end play of 0.001-0.004. Then, the worm helix width should be adjusted to provide the valve actuating sleeve with an end play of 0.0015-0.0025. To make the adjustments, proceed as follows:

Remove grille, disconnect steering shaft and pull the steering shaft rearward and free from the valve actuating sleeve. Remove the large plug (22—Fig. JD1775A), install a cap screw in end of steering worm and mount a dial indicator as shown in Fig. JD1775B. Rotate the valve actuating sleeve assembly in both directions and note the amount of end play in the worm bearings as registered on the dial indicator. If the indicator reading is not 0.001-0.004, add or deduct shims

(19—Fig. JD1775A) until proper end play is obtained. It is important that end play be adjusted as close to 0.001 as possible.

With the worm shaft end play properly adjusted, mount a dial indicator in a suitable manner, similar to that shown in Fig. JD1775C, reach into the large screw plug opening in front bearing housing with your fingers and while pressing on the worm to keep it from moving, move valve actuating sleeve back and forth to determine amount of end play between the hardened dowel and the helix. If the indicator reading is not 0.0015-0.0025, adjust the helix width by loosening the special lock screw (15—Fig. JD1775A), which is located inside bushing (14), and using a ⅞-inch socket, turn the bushing **in** to decrease or **out** to increase the end

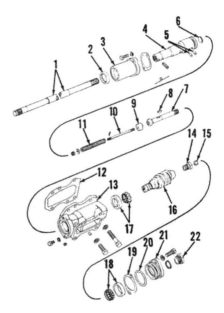

Fig. JD1774 — Exploded view of typical early production 520, 620 and 720 power steering worm, valve actuating sleeve and associated parts. Backlash between the worm and steering gear is controlled by shims (65).

43. Woodruff key
44. Steering shaft
45. Oil seal
46. Oil seal housing
47. Gasket
48. Valve actuating sleeve
49. Pins
50. Expansion plug
51. Cone point set screw
52. Bearing housing
53. "O" ring
54. Shims
55. Bearing cup
56. Bearing cone
57. Steering worm
58. Worm cam
59. Cam rod
60. Cotter pin
61. Spring
62. Washer
63. Nut
64. Worm housing
65. Shims
66. Pipe plug
67. Dowel pin

Fig. JD1775A — Exploded view of late production 520, 620 and 720 steering worm, valve actuating sleeve and associated parts. Backlash between the worm and steering gear is controlled by shims (12). Series 530, 630 and 730 are similar.

4. Actuating sleeve	15. Lock screw
5. Dowels	16. Worm
6. Welch plug	17. Bearing cup and cone
7. Worm shaft	
8. Woodruff key	18. Bearing cup and cone
9. Cam	
10. Spring rod	19. Adjusting shim
11. Cam spring	20. "O" ring
12. Adjusting shims	21. Worm bearing housing
13. Worm housing	
14. Adjusting bushing	22. Screw plug

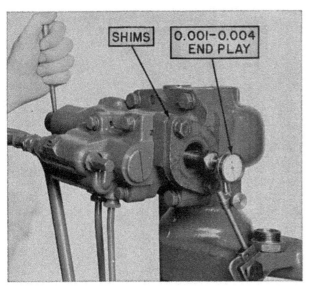

Fig. JD1775B—Using dial indicator to check the steering worm end play on late production series 520, 620 and 720. Series 530, 630 and 730 are similar.

play clearance. Adjust the clearance as close to 0.0015 as possible. When adjustment is complete, hold the bushing from turning and tighten the lock screw.

Note: Never make the adjustment so tight that steering wheel will not return to neutral when released after being turned in either direction to operate the steering valve. Refer to paragraph 25A. A special tool for adjusting the helix width is available from R. B. Precision Tools, 313 Morse St., Ionia, Mich.

26A. STEERING SHAFT BIND. To check and/or adjust the steering shaft bind, refer to paragraph 25A.

26B. BACKLASH. To check and/or adjust the backlash between the steering worm and gear, refer to paragraph 25B.

26C. R&R AND OVERHAUL. To remove the steering worm and valve actuating sleeve, first remove the grille, disconnect the steering shaft coupling and drain the power steering oil reservoir and worm housing. Disconnect the oil lines and remove the steering valve housing as outlined in paragraph 24. Unbolt and remove the worm shaft housing from pedestal, being careful not to damage or lose the backlash adjusting shims. Remove

the worm shaft front bearing housing (21—Fig. JD1775A) and carefully measure the distance the adjusting bushing protrudes from the worm as shown in Fig. JD1775D. Loosen the special lock screw, remove the adjusting bushing and withdraw the worm.

Remove oil seal housing (3—Fig. JD1775A) and pull actuating sleeve and worm shaft assembly from housing. To remove the worm shaft from the actuating sleeve, use a pair of vise-grip pliers to turn the shaft about 90 degrees in a counter-clockwise direction so the cam and shaft keyways align.

Clean and inspect all parts and renew any that are damaged and cannot be reconditioned. Lip of oil seal (2) goes toward worm. Renew worm if it is worn or if helix end is scored. Renew worm shaft (7) if it is worn or scored at cam or helix locations. Small burrs can be removed from wearing surfaces by using a fine stone. Check the valve actuating sleeve dowels (5) for damage, and bore of sleeve for

being scored. Check fit of parts by inserting them in the sleeve. If binding exists, remove the dowel pins and hone the sleeve bore to insure a free fit. Renew the dowel pins if they are worn or damaged.

When reassembling, use a valve spring tester and measure the length of spring (11) at 70-80 lbs. pressure. Note: If the spring will not exert 70-80 pounds pressure at $1\frac{7}{8}$ inches or more, obtain a new spring and measure its length at 70-80 lbs. pressure. Then install spring on spring rod and tighten the adjusting nut until spring is compressed to the same length which yielded a pressure of 70-80 lbs.

Assemble worm shaft (7) and cam (9) with keyways aligned. Then insert the worm shaft assembly into the actuating sleeve so the keyways will slide under the dowels and when the cam is behind the front dowel, turn the worm shaft clockwise with a pair of vise-grip pliers until the cam seats. Install actuating sleeve and worm shaft assembly into worm housing and install bearing assembly (17). With Woodruff key properly positioned in worm shaft, slide worm into position.

Using a small screw driver as shown in Fig. JD1775E, move the worm shaft until it lacks about 1/32-inch of being flush with end of worm; then, carefully install the adjusting bushing and screw it into the worm until all actuating sleeve end play is removed (dowel tight in helix). Measure the bushing protrusion as shown in Fig. JD1775D. If the measured dimension is not almost identical to that measured before the bushing was removed, unscrew the bushing and start over.

When proper bushing protrusion is obtained, install lock screw, front bearing housing and oil seal housing. Check the end play adjustments as in paragraph 26.

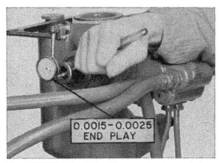

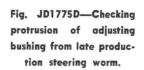

Fig. JD1775C—Checking end play of late production 520, 620 and 720 steering valve actuating sleeve. Specified end play is 0.0015-0.0025. Series 530, 630 and 730 are similar.

Fig. JD1775D—Checking protrusion of adjusting bushing from late production steering worm.

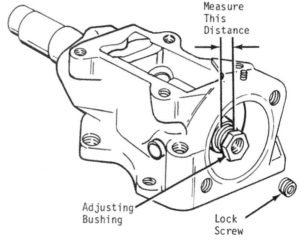

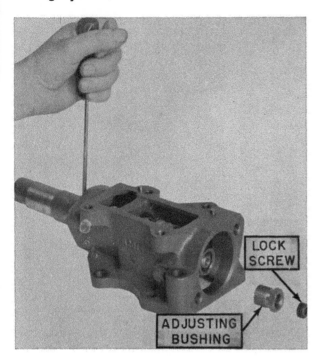

Fig. JD1775E — Positioning the late production worm shaft with respect to the worm prior to installing adjusting bushing.

Install the worm housing assembly and adjust the gear backlash as in paragraph 25B. Install the steering valve housing as in paragraph 24A.

SPINDLE, PEDESTAL AND CYLINDER

29B. R&R AND OVERHAUL. Remove the grille, disconnect the steering shaft coupling and drain the power steering reservoir. Disconnect the oil lines, unbolt worm housing from pedestal and remove the worm housing and valve housing assembly being careful not to lose the backlash adjusting shims located between the worm housing and pedestal. Using a feeler gage as shown in Fig. JD1776, measure the spindle end play clearance between the steering gear and the adjusting shims. If the clearance is not within the limits of 0.004-0.021, it will be necessary to add or remove the necessary amount of shims during reassembly. Remove the cup plug

from top of pedestal and the cap screw and washer retaining the steering gear to vertical spindle. Using a wood block or brass drift and a heavy hammer, drive the steering gear upward until gear is loose on the spindle splines. Unbolt pedestal from front end support and lift pedestal, gear and adjusting shims from spindle.

Remove cylinder (76—Fig. JD1777) and unbolt wheel fork, lower spindle or center steering arm from the steering spindle (83). Unlock the cap screws (77) and unbolt the vane bracket from spindle. Lift the quill (92) and steering spindle assembly from tractor and bump spindle out of quill.

Lower surface of pedestal and upper surface of quill (92) form part of the cylinder sealing surface and must be smooth and free from nicks and scratches. Original inside diameter of the pedestal upper bushing (70) is 1.502-1.503. When installing the bushing, the beveled edge should be toward bottom of pedestal, the split should be toward the steering worm and top of bushing should be flush with top of bushing bore. New bushing is pre-sized and, if not damaged during installation, will require no

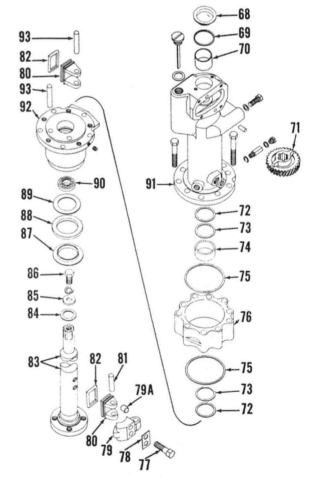

Fig. JD1777 — Exploded view of power steering pedestal, steering spindle and associated cylinder parts.

68. Cup plug
69. Gasket
70. Bushing
71. Steering gear
72. Back-up gasket
73. "O" ring
74. Bushing
75. Gasket
76. Steering cylinder
77. Special cap screws
78. Lock plate
79. Vane bracket
79A. Hollow dowel (late model)
80. Steering vane
81. Vane pin
82. Vane packing
83. Steering spindle
84. Shim washers
85. Washer
86. Cap screw
87. Retainer
88. Cork seal
89. Washer
90. Thrust bearing
91. Pedestal
92. Spindle quill
93. Vane pin

Fig. JD1776 — Checking the power steering spindle end play. Desired end play of 0.004-0.021 is controlled by shims under the gear.

final sizing. Original inside diameter of the pedestal lower bushing (74) is 2.502-2.503. To remove the bushing, extract the "O" ring located just above the bushing and use a suitable puller. Using a closely fitting mandrel, install bushing until it is flush with bottom of pedestal. Bushing is not pre-sized and must be honed to an inside diameter of 2.502-2.503. After honing, clean and lubricate the bushing. Install a new "O" ring and back-up washer above the lower bushing. Note: The upper end of the bushing forms the lower edge of the "O" ring groove. Install a new and well lubricated neoprene gasket in groove in lower flange of pedestal.

Bearing surface of spindle must be smooth and free from nicks or scratches. Side of thrust bearing (90) marked "THRUST HERE" must be installed toward bottom of quill (92). Install washer (89), new cork seal (88) and retainer (87). Install back-up washer (72) in quill, then install a new "O" ring on top of the washer. With "O" ring (73) well lubricated, slide spindle into quill, position the assembly on the front end support and bolt wheel fork, lower spindle or center steering arm in position and tighten the cap screws to a torque of 275 ft.-lbs.

Install the steering vane bracket (79), and use a 0.005 thick shim (John Deere Part No. R21231R) between vane bracket and spindle. This shim should be centered between the locating dowels on late models. Tighten the bracket retaining cap screws to a torque of 275 ft.-lbs. on models with dowels or 208 ft.-lbs. on models without the hollow dowels. Bend tab of lock plate over cap screw heads. Install vane seals, vanes and pins. Refer to Fig. JD1778. Install cylinder (76—Fig. JD1777) and position the cylinder so that all cap screw holes in cylinder and quill are aligned. Lubricate "O" ring (73) in lower part of pedestal and install pedestal. When lowering pedestal into position, install the end play adjusting shims (84) and steering gear. Tighten gear retaining cap screw to a torque of 192 ft.-lbs.

Note: The number of end play adjusting shims to be inserted between steering gear and pedestal were determined during disassembly and should provide the steering spindle with an end play of 0.004-0.021 when the pedestal retaining cap screws are securely tightened.

Beginning with the front center cap screw hole in pedestal, install a long cap screw in every other hole. Install the remaining shorter cap screws and tighten all of the pedestal cap screws to a torque of 150 ft.-lbs.

Assemble the remaining parts and adjust the steering gear backlash as in paragraph 25B. Remove any binding in the steering shaft as outlined in paragraph 25A.

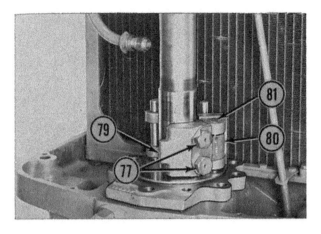

Fig. JD1778 — Correct installation of the steering vane bracket and vane. Screws (77) should be tightened to a torque of 208 Ft.-Lbs.

POWER STEERING SYSTEM
(Series 620 Orchard)

Note: The maintenance of absolute cleanliness of all parts is of utmost importance in the operation and servicing of the hydraulic power steering system. Of equal importance is the avoidance of nicks or burrs on any of the working parts.

LUBRICATION AND BLEEDING

29C. Working fluid for the power steering system is supplied by the hydraulic "Powr-Trol" pump via a flow control valve. To drain, refill and bleed the system, refer to the "Powr-Trol" section.

SYSTEM OPERATING PRESSURE

29D. The system relief valve is mounted in the flow control valve housing as shown in Fig. JD1780. To check the system operating pressure, connect a suitable pressure gage and shut-off valve in series with the pressure line connecting the flow control valve housing to the cylinder and valve unit in a manner similar to that shown in Fig. JD1779. Notice that the shut-off valve is connected in the circuit between the gage and the steering valve. Open the shut-off valve, engage the "Powr-Trol" pump and run the engine at low idle speed until the oil is warmed. Advance the engine speed, close the shut-off valve, and observe the pressure gage reading; then open the shut-off valve immediately. If the gage reading is approximately 1000 psi with the shut-off valve closed, the "Powr-Trol" pump and relief valve are O.K. and any trouble is located elsewhere in the system.

If the pressure reading is not as specified, remove the relief valve plug (8—Fig. JD1780) and vary the number

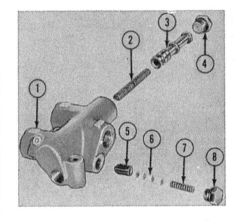

Fig. JD1780—Exploded view of the flow control and relief valve used in the power steering system of series 620 orchard.

1. Body	4. Plug
2. Flow control valve spring	5. Relief valve
	6. Shim washers
3. Flow control valve	7. Relief valve spring
	8. Plug

Fig. JD1779 — Shut-off valve and pressure gage installation diagram for trouble shooting the power steering system on series 620 orchard.

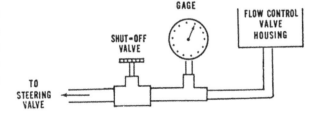

of shim washers (6) as required and recheck the pressure reading. If the pressure cannot be increased to approximately 1000 psi, and the flow control valve is known to be in good condition, check for a faulty hydraulic pump.

PUMP

29E. Working fluid for the power steering system is supplied by the hydraulic "Powr-Trol" pump. To overhaul the pump, refer to the "Powr-Trol" section.

FLOW CONTROL AND RELIEF VALVE

29F. OVERHAUL. With the unit removed from tractor, disassemble the unit as shown in Fig. JD1780, thoroughly clean all parts and renew any which are damaged or excessively worn. Renew the nylon relief valve (5) if it shows any roughness. If the relief valve seat in housing is not excessively worn, it can be reconditioned by lapping. If a normal amount of lapping will not restore the seat, renew the housing (1). Renew the flow control valve (3) and spring (2) if their condition is questionable.

When reassembling, tighten plugs (4 & 8) securely. After unit is installed on tractor, check the system operating pressure as outlined in paragraph 29D.

STEERING CYLINDER AND VALVE

29G. R&R AND OVERHAUL. To remove the unit, disconnect the oil lines and control rods at cylinder, disconnect the drag link at steering knuckle arm and loosen the clamp screws at front end of cylinder and rear end of piston rod. Turn drag link out of drag link support and turn piston rod out of rod end socket. Note: If piston rod cannot be easily unscrewed from the socket, disconnect socket (end) from arm.

With the cylinder and valve unit removed, drain the oil and remove the cap screw and the five self-locking nuts from valve end of tie rods. Scribe marks on the drag link support (21—

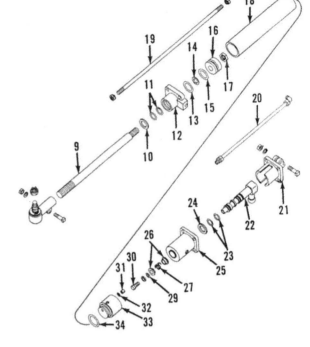

Fig. JD1782 — Exploded view of the power steering cylinder and valve unit used on series 620 orchard tractors.

  9. Piston rod
10. Oil seal
11. "O" ring packing
12. Piston rod guide
13. "O" ring packing
14. Gasket, copper
15. "O" ring packing
16. Piston
17. Jam nut
18. Cylinder
19. Tie rod (4 used)
20. Oil pipe
21. Drag link support
22. Piston rod valve
23. "O" ring packing
24. Oil seal
25. Valve housing
26. Retainer
27. Spring
29. "O" ring packing
30. Screw
31. Internal tube
32. "O" ring packing
33. Valve cap
34. "O" ring packing

Fig. JD1782) and valve housing (25) to facilitate assembly in the correct position, then disconnect oil pipe (20) from valve housing and, using a soft hammer, tap the four tie rods until they are free of the valve housing. Remove drag link support and valve housing from cylinder. Clamp the round head of centering spring retainer screw (30) in vise grip pliers and remove the screw. Remove centering spring and retainer assembly (26 & 27) and withdraw valve (22) from housing.

Withdraw valve cap (33) from cylinder, remove nut (17) and withdraw piston (16). Extract copper gasket (14) and withdraw the piston rod. The need and procedure for further disassembly is evident.

Thoroughly clean all parts and renew any which are damaged or worn beyond repair. The valve and housing are available as a complete assembly only.

When reassembling, be sure to renew all "O" rings and seals and proceed as follows: Lubricate the piston rod and slide the piston rod guide onto rod; then lubricate the cylinder and assemble cylinder to piston rod guide. Install copper gasket (14), then install piston so that the flat machined surface of same faces inside (rear) of cylinder. Tighten the piston retaining nut securely and stake same in place. Install valve cap (33). Install oil seal on valve (22), then install "O" ring on ball end of valve. Push valve completely into valve housing and install the other "O" ring on the protruding end of valve; then, drive the oil seal in until flush with outer face of housing. Be sure to rotate valve as seal is installed to prevent cocking the seal. Install centering spring and retainers, then install and securely tighten the round head screw (30).

Attach oil line to piston rod guide, align scribe marks and install valve assembly and drag link support. Connect oil line to valve housing, install tie rods and tighten the nuts securely. Screw drag link into drag link support and rod end socket onto piston rod. The overall length of the cylinder assembly when fully retracted should be $27\frac{9}{16}$-$27\frac{11}{16}$ inches. The overall length of the cylinder assembly with drag link installed should be $48\frac{1}{16}$-$48\frac{3}{16}$ inches.

Install the assembly on tractor, connect oil lines and refill the "Powr-Trol" reservoir. Overall length of steering cylinder control rod should be $33\frac{9}{16}$-$33\frac{11}{16}$ inches.

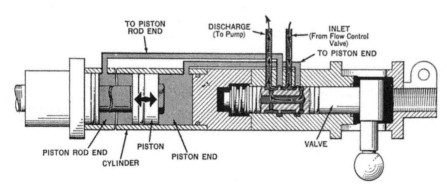

Fig. JD1781—Sectional view of the power steering cylinder and valve unit used on series 620 orchard tractors. Working fluid for the power steering system is supplied by the "Powr-Trol" pump.

ENGINE AND COMPONENTS

The engine crankcase and the tractor main case is an integral unit. A wall in the main case separates the crankcase compartment from the transmission and differential compartment.

CYLINDER HEAD
Series 520-530-620-630

30. **REMOVE AND REINSTALL.** To remove the cylinder head, first drain cooling system, loosen hose clamps and remove the water inlet casting (lower water pipe) from cylinder head. Disconnect choke and throttle rod from carburetor. Disconnect fuel line and remove carburetor, exhaust pipe and manifold.

Remove tool box, ventilator tube and tappet lever cover. Remove rocker arms assembly and withdraw push rods. Remove the cylinder head retaining stud nuts, slide cylinder head forward on studs and withdraw head from tractor.

When reassembling, install gasket dry with smooth (steel) side toward block and permatex the mating surfaces of gasket and cylinder block. Using new lead washers under the stud nuts, install the nuts and tighten them progressively from center outward and to a torque of 104 ft.-lbs. for series 520 and 530, 150 ft.-lbs. for series 620 and 630. Reinstall tappet lever assembly making certain that tappet lever supports engage the positioning

dowels in head and adjust tappets to 0.022 cold.

After engine is hot, re-tighten the head nuts to the specified torque value and adjust the tappet gap to 0.020 hot.

Series 720-730

31. **REMOVE AND REINSTALL.** To remove the cylinder head, first drain cooling system, unbolt the water inlet casting from head and remove casting and lower water pipe. Disconnect choke and throttle rods from carburetor. Disconnect fuel line and remove carburetor and lower half of air intake elbow. Remove exhaust pipe, unbolt and remove manifold. Remove ventilator tube and tappet lever cover. Remove rocker arms assembly and withdraw push rods. Remove the cylinder head retaining stud nuts, slide cylinder head forward on studs and withdraw head from tractor.

When reassembling, install gasket dry with smooth (steel) side toward block and permatex the mating surfaces of gasket and cylinder block. Using new lead washers under the stud nuts, install the nuts and tighten them progressively from center outward and to a torque of 275 ft.-lbs. Install tappet levers assembly making certain that tappet lever supports engage the positioning dowels in head and adjust tappets to 0.022 cold.

After engine is hot, re-tighten the

head nuts to the specified torque value and adjust the tappet gap to 0.020 hot.

VALVES AND SEATS
All Models (Except LP-Gas)

32. Intake and exhaust valves are not interchangeable. Valves seat directly in cylinder head with a seat angle as follows:

Intake

All-fuel engines ....30°
Gasoline engines ....45°

Exhaust, All models...........45°

Valve face angle is as follows:

Intake

520-530-620-630 All-fuel .....29¾°
720-730 All-fuel ....29½°
Gasoline engines ....44½°

Exhaust, All models...........44½°

Desired seat width is as follows:

Intake:

520-530-620-630 ....⅛-9/64 inch
720-730 All-fuel ....5/32-11/64 inch
720-730 Gasoline ......⅛-9/64 inch

Exhaust:

All models ....⅛ inch

Seats can be narrowed, using 20 and 70 degree stones.

Intake and exhaust valve stem diameter is as follows:
520-530 ....0.3715-0.3725
620-630-720-730 ....0.496 -0.497

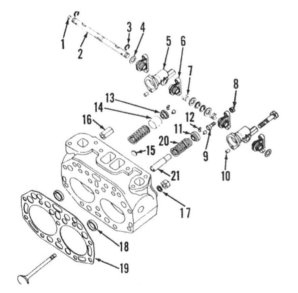

Fig. JD1783—Exploded view of 520 and 530 cylinder head, rocker arms and valve components.

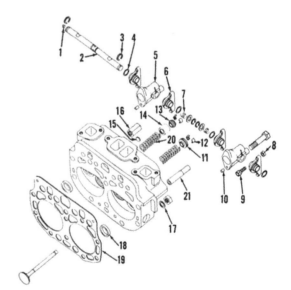

Fig. JD1784—Exploded view of 720 and 730 cylinder head, rocker arms and valve components. Series 620 and 630 are similar.

1. Plug	7. Spring	10. Dowel pin
2. Rocker arm shaft	8. Jam nut	11. Exhaust valve rotator
3. Snap ring	9. Tappet adjusting screw	12. Keepers
4. Washers		13. Intake valve spring cap
5. Bracket		
6. Rocker arm		

14. Oil shield	18. Exhaust valve seat insert (LP-Gas only)
15. Expansion plug	19. Head gasket
16. Coupling nut	20. Valve spring
17. Lead washer	21. Valve guide

Adjust intake and exhaust tappet gap to 0.020 hot. If gap is adjusted with engine cold, allow 0.002 for expansion purposes and recheck the gap when engine is hot.

Refer to paragraph 35 for valve rotator information.

LP-Gas Models

33. Intake valves are not interchangeable with the exhaust valves which are equipped with positive type rotators. Refer to paragraph 35 for valve rotator information. Intake valves seat directly in cylinder head with a face angle of 44½ degrees and a seat angle of 45 degrees. Exhaust valves seat on renewable, alloy seat inserts with a face angle of 44½ degrees and a seat angle of 45 degrees. Desired seat width is as follows:

Intake:
520-530 .... ⅛-9/64 inch
620-630-720-730 .... ⅛-9/64 inch

Exhaust:
520-530 .... 7/64-⅛ inch
620-630-720-730 .... ⅛-9/64 inch

Seats can be narrowed using 20 and 70 degree stones.

Refer to paragraph 34 for method of renewing the valve seat inserts. Intake and exhaust valve stem diameter is 0.496-0.497 for series 620, 630, 720 and 730, 0.3715-0.3725 for series 520 & 530.

Adjust intake and exhaust tappet gap to 0.020 hot. If gap is adjusted with engine cold, allow 0.002 for expansion purposes and recheck the gap when engine is hot.

34. Exhaust valve seat inserts can be removed by using a suitable puller. New inserts are available in standard size only and the head is reamed to provide an interference fit. Thoroughly clean the counterbore in cylinder head before attempting to install a new seat. Chill seat in dry ice to facilitate installation.

After installation, check the concentricity of the seat with respect to the valve guides. Seats should be concentric within 0.002.

VALVE ROTATORS
All Models

35. Positive type exhaust valve rotators (Fig. JD1785) are factory installed on all models. The rotators can be considered in satisfactory operating condition if the exhaust valves rotate a slight amount each time the valves open. Servicing consists of renewing the complete rotator unit.

VALVE GUIDES AND SPRINGS
All Models

36. Intake and exhaust valve guides are interchangeable in any one model and can be driven from cylinder head if renewal is required. Press new guides into cylinder head so that smaller O.D. of guides will be toward valve springs. Distance from port end of guides to gasket surface of cylinder head is as follows:

Intake:
Series 520-530 .... 2 13/16 inches
Series 620-630 .... 3 1/16 inches
Series 720-730 .... 3 9/32 inches

Exhaust:
Series 520-530 .... 2 11/16 inches
Series 620-630 .... 2 23/32 inches
Series 720-730 .... 3 9/32 inches

Ream new guides after installation to the following inside diameter:
Series 520-530 .... 0.3759-0.3775
Series 620-630-720-730 0.5009-0.5025
Recommended clearance between valve stems and guides is as follows:
Series 520-530 .... 0.0034-0.0060
Series 620-630-720-730 0.0039-0.0065

37. Intake and exhaust valve springs are interchangeable in any one model. Renew any spring which is rusted, discolored or does not meet the load test specifications which follow:

Pounds Test @ Height in Inches:
Series 520-530 ..... 30.5-36.5 @ 2 13/32
Series 620-630 .... 36-44 @ 2 13/32
Series 720-730 ..... 45-55 @ 3 7/64

VALVE TAPPET LEVERS
(Rocker Arms)
All Models

38. **R&R AND OVERHAUL.** The procedure for removing the rocker arms assembly is evident after removing the rocker arm cover. Check the tappet levers and shaft against the values which follow:

Shaft Diameter
Series 520-530 .... 0.672-0.673
Series 620-630-720-730.. 0.858-0.859

Tappet Lever Bore:
Series 520-530 .... 0.676-0.678
Series 620-630-720-730.. 0.861-0.863

Shaft Clearance In Lever:
Series 520-530 .... 0.003-0.006
Series 620-630-720-730.. 0.002-0.005

Excessive wear of any of the component parts of the tappet lever assembly is corrected by renewing the parts. Intake and exhaust tappet levers are interchangeable in any one model.

When reinstalling, make certain that tappet lever supports engage the positioning dowels in cylinder head and adjust tappet gap to 0.020 hot.

VALVE TIMING
All Models

39. Valves are properly timed when "V" marked tooth space on camshaft gear is in register with the "V" marked tooth on the crankshaft gear as shown in Fig. JD1786. The preferred and probably the quickest method of checking the timing is to remove the crankcase cover and observe the timing marks.

Valve timing, however, can be checked without removing crankcase cover, as follows:

First make certain that valve tappet gap is adjusted to 0.020 hot. Turn crankshaft until the exhaust valve of No. 1 (LEFT) cylinder is just beginning to open (0.000 tappet gap). At this time, the flywheel mark "LHEO"

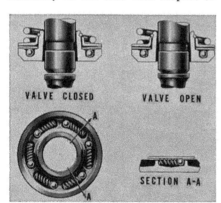

Fig. JD1785—Positive type exhaust valve rotators are factory installed on all models. Servicing of the rotator consists of renewing the complete unit.

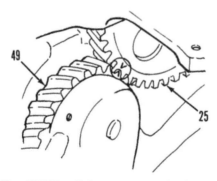

Fig. JD1786 — Valves are properly timed when "V" marked tooth space on camshaft gear (25) is meshed with "V" marked tooth on crankshaft gear (49). The preferred method of checking the valve timing is to remove the crankcase cover and observe the marks.

Fig. JD1787—Valve timing mark "LHEO" is located on the outer rim of the flywheel. The mark can be viewed through the flywheel housing inspection port as shown.

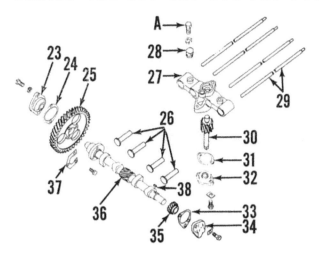

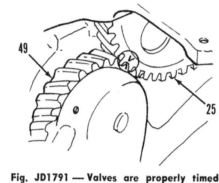

Fig. JD1788 — 520 and 530 camshaft, cam gear, cam followers and associated parts. Engine oil pump is driven by gear and shaft (30).

23. Camshaft left bearing
24. Shim gaskets
25. Cam gear
26. Cam followers
27. Cam follower guide
28. Follower guide support
29. Push rods
30. Oil pump drive gear
31. Gasket
32. Oil pump drive gear bearing
33. Gasket
34. Camshaft right bearing
35. Distributor drive gear
36. Camshaft
37. Lock plate
38. Woodruff key

Fig. JD1791 — Valves are properly timed when "V" marks on camshaft gear and crankshaft gear are in register as shown.

should be approximately in line with notch in the timing hole in flywheel cover as shown in Fig. JD1787.

TIMING GEARS
All Models

40. The procedure for renewing the camshaft gear and/or crankshaft gear is outlined in the paragraphs which cover the renewal of the respective shafts.

CAMSHAFT AND BEARINGS
Series 520-530

45. R&R AND OVERHAUL. To remove the camshaft gear and/or shaft and bearings, first remove governor and housing assembly as outlined in paragraph 102, clutch and belt pulley as in paragraph 118 and the reduction gear cover as outlined in paragraph 123. Remove the crankcase cover, rocker arm cover, rocker arms assembly and push rods. Remove ignition distributor assembly. Remove the camshaft right and left bearings and completely unscrew the three cap screws (A—Figs. JD1788 & 1789) which retain the cam follower guide to the main case. Straighten tabs on locking plate (37) and unbolt gear from shaft. Buck up the camshaft gear and bump

camshaft toward right and out of gear. The camshaft gear can be removed from above at this time.

With the camshaft gear removed, disconnect both of the oil lines which connect to top of inside of crankcase. Move camshaft toward right and withdraw shaft through crankcase cover opening.

The 0.7505-0.7515 diameter camshaft bearing journals have a normal operating clearance of 0.0015-0.0045 in the 0.753-0.755 diameter cast iron bearings. If the running clearance is excessive, renew the shaft and/or bearings.

When reassembling, chamfer (C—Fig. JD1790) on camshaft gear must be toward shoulder on camshaft. Install gear on camshaft so that when right hand lobe (L) is pointing upward, the "V" mark on gear is at the top. Bolt gear to camshaft and lock the cap screws in position by upsetting tabs of lock plate (37—Fig. JD1788).

Mesh camshaft gear with crankshaft gear so that timing marks are in register as shown in Fig. JD1791 and vary the number of shims (24—Fig. JD1788) which are located between the left bearing and the main case to give camshaft an end play of 0.005-0.012 when checked with a dial indicator as shown in Fig. JD1792. Tighten the camshaft bearing retaining screws to a torque of 35 Ft.-Lbs. Install the remaining parts by reversing the disassembly procedure, check the ignition timing

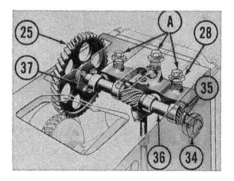

Fig. JD1789—520 and 530 camshaft installation. The cam follower guide is retained by cap screws (A).

Fig. JD1790 — When installing camshaft gear on 520 and 530, chamfer (C) must be toward shoulder on shaft and right lobe (L) must point in same direction as "V" mark on gear.

Fig. JD1792 — Using a dial indicator to check 520 and 530 camshaft end play. Desired end play of 0.005-0.012 is controlled by shims (24) located under bearing (23).

as in paragraph 113 and adjust the valve tappet gap to 0.020 hot.

Series 620-630-720-730

46. R&R AND OVERHAUL. To remove the camshaft gear and/or shaft and bearings, first remove the flywheel as outlined in paragraph 59, governor as in paragraph 103, clutch and belt pulley as in paragraph 118 for 620 & 630 or 119 for 720 & 730 and the reduction gear cover as outlined in paragraph 130 for 620 & 630 or 136 for 720 & 730. Remove the crankcase cover, rocker arm cover,

rocker arms assembly and push rods. Remove the ignition distributor assembly. Remove the camshaft left bearing housing (39—Fig. JD1793) and using a suitable puller as shown in Fig. JD1794, remove the shaft left bearing cone. Straighten tabs on locking plate (50 or 50A—Fig. JD1793), and unbolt gear from shaft. Buck up the camshaft gear and bump camshaft toward right and out of gear. The camshaft gear can be removed from above at this time.

With the camshaft gear removed, disconnect both of the oil lines which connect to top of inside of crankcase. Move camshaft toward right and withdraw shaft through crankcase cover opening.

When reassembling, bolt gear to camshaft and lock the cap screws in place by upsetting tabs of lock plate (50 or 50A). Mesh camshaft gear with crankshaft gear so that timing marks are in register as shown in Fig. JD1791 and install the left bearing housing. Camshaft end play is controlled by a spring at right end of shaft. Make certain that the spring is in position when

installing the reduction gear cover. Install the remaining parts by reversing the disassembly procedure, check the ignition timing as outlined in paragraph 113 and adjust the valve tappet gap to 0.020 hot.

CAM FOLLOWERS
Series 520-530

48. The mushroom type cam followers (26—Fig. JD1788) operate directly in the unbushed bores of follower guide (27). To remove the followers and guide assembly, first remove camshaft as outlined in paragraph 45. Remove the oil pump cover and idler gear. Grasp drive gear and pull drive gear and shaft down and out of pump body. Refer to Fig. JD1795. Withdraw follower guide and followers from above.

Excessive wear between the cam followers and the guide is corrected by renewal of the component parts.

Series 620-630-720-730

49. The mushroom type cam followers (47—Fig. JD1793) operate directly in machined bores in the tractor main case. To remove the followers, remove

camshaft as outlined in paragraph 46 and withdraw followers from the case bores. Excessive clearance between the followers and the case bores is corrected by renewal of the parts.

ROD AND PISTON UNITS
All Models

50. To remove the connecting rod and piston units, first remove cylinder head as outlined in paragraphs 30 or 31. Remove the "Powr-Trol" pump. Remove the crankcase cover, connecting rod caps, bearing inserts and rod bolts. Remove carbon accumulation and ridge from unworn portion of cylinders to prevent damage to ring lands and to facilitate piston removal.

Using a piece of 2x4 wood, make up a set of blocks shown in Fig. JD1796. Using the smallest block between connecting rod and crankshaft, turn crankshaft and push rod and piston unit forward. Continue this process with the next largest block, and so on, until pistons can be withdrawn from front.

When reinstalling, numbers on rod and cap should be in register and face toward top of engine. Number one cylinder is on left side of tractor. Tighten rod bolts to a torque 85 Ft.-Lbs. for series 520, 530, 720 and 730; 105 Ft.-Lbs. for series 620 and 630.

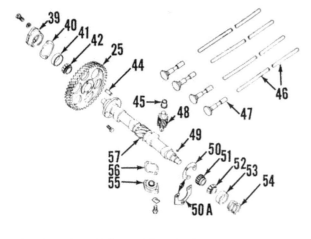

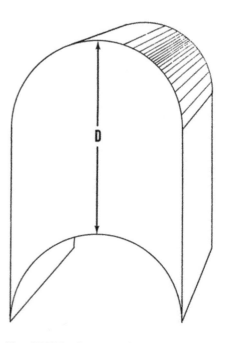

Fig. JD1793 — 620, 630, 720 and 730 camshaft, cam gear, cam followers and associated parts. The ignition distributor is driven by gear (51).

25. Cam gear
39. Left bearing housing
40. Gasket
41. Bearing cup
42. Bearing cone
44. Dowel pin
45. Bushing
46. Push rods
47. Cam followers
48. Oil pump drive gear
49. Woodruff key
50. Lock plate (620 & 630)
50A. Lock plate (720 & 730)
51. Distributor drive gear
52. Bearing cone
53. Bearing cup
54. Spring
55. Oil pump drive gear bearing
56. Gasket
57. Camshaft

Fig. JD1794—Using a special puller to remove the camshaft left bearing cone. This procedure is typical of that used on series 620, 630, 720 and 730.

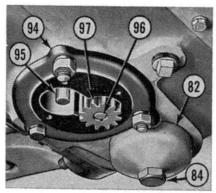

Fig. JD1795—Bottom view of typical main case, showing the oil pump and filter installation. The oil pump cover and idler gear have been removed.

Fig. JD1796—Suggested home made wood block which can be used to push connecting rod and piston units forward, out of cylinder block. It is more convenient to use three blocks. Dimension (D) should be approximately as follows: 1st. block, 2⅞ inches; 2nd. block, 4 inches; 3rd. block, 5⅝ inches. Refer to text.

PISTONS AND RINGS
All Models

51. Pistons and rings are available in standard as well as 0.045 oversize for all models. Rings should be installed with dot toward head (closed end) of piston.

Inspect pistons, rings and cylinder walls using the values which follow:

Series 520-530
Cylinder bore, Std. 4.6871-4.6885
Piston diameter at top
of skirt just above
piston pin
    Parallel to piston pin. .4.6736-4.6746
    Right angles to
        piston pin 4.6806-4.6816
Piston diameter at
bottom of skirt
    Parallel to piston pin. .4.6746-4.6756
    Right angles to
        piston pin 4.6816-4.6826
Piston skirt clearance at
top of skirt just above
piston pin
    Parallel to piston pin. .0.0125-0.0149
    Right angles to
        piston pin 0.0055-0.0079
Piston skirt clearance at
bottom of skirt
    Parallel to piston pin. .0.0115-0.0139
    Right angles to
        piston pin 0.0045-0.0069
Piston ring side clearance
    No. 1 (front of piston) . . .0.004-0.006
    Nos. 2 & 3. . . . . .0.004-0.006
    Nos. 4 & 5. . . . . .0.0045-0.0075
    No. 6 . . . . .0.0035-0.0055

Series 620-630
Cylinder bore, Standard. .5.4996-5.5010
Piston diameter at top
of skirt just above
piston pin
    Parallel to piston pin. .5.4923-5.4933
    Right angles to
        piston pin 5.4930-5.4940
Piston diameter at
bottom of skirt
    Parallel to piston pin. .5.4933-5.4943
    Right angles to
        piston pin 5.4940-5.4950
Piston skirt clearance at
top of skirt just above
piston pin
    Parallel to piston pin. .0.0063-0.0087
    Right angles to
        piston pin 0.0056-0.0080
Piston skirt clearance at
bottom of skirt
    Parallel to piston pin. .0.0053-0.0077
    Right angles to
        piston pin 0.0046-0.0070
Piston ring side clearance
    No. 1 (front of piston). . . .0.004-0.006
    Nos. 2 & 3. . . . . .0.004-0.006
    Nos. 4 & 5. . . . . .0.0045-0.0075
    No. 6 . . . . .0.0035-0.0055

Series 720-730
Cylinder bore, Standard. .5.9996-6.0010
Piston diameter at top of
skirt just above piston pin
    Parallel to piston pin. .5.9835-5.9845
    Right angles to
        piston pin 5.9905-5.9915
Piston diameter at bottom
of skirt
    Parallel to piston pin. .5.9845-5.9855
    Right angles to
        piston pin 5.9915-5.9925
Piston skirt clearance at
top of skirt just above
piston pin
    Right angles to
        piston pin 0.0081-0.0105
Piston skirt clearance at
bottom of skirt
    Right angles to
        piston pin 0.0071-0.0095
Piston ring side clearance
    No. 1 (front of piston) . . .0.004-0.006
    Nos. 2 & 3. . . . . .0.004-0.006
    Nos. 4 & 5. . . . . .0.0045-0.0075
    No. 6 . . . . .0.004-0.006
    No. 7 . . . . .0.0035-0.0055

PISTON PINS AND BUSHINGS
All Models

52. The full floating type piston pins are retained in the piston pin bosses by snap rings and are available in standard as well as oversizes of 0.003 (marked yellow) and 0.005 (marked red).

Piston Pin Diameter is as Follows:
    Series 520-530 . . . . .1.4166-1.4170
    Series 620-630 . . . . . . .1.7495-1.7500
    Series 720-730 . . . . . . .2.3545-2.3550
    Piston pins should have 0.000-0.001 clearance in piston and 0.001-0.0024 clearance in connecting rod bushing.

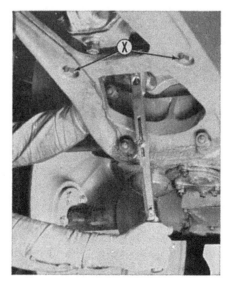

Fig. JD1797 — Tightening stud nuts which retain cylinder block to the main case. Cap screws (X) secure block to front end support.

CONNECTING RODS AND BEARINGS
All Models

53. Connecting rod bearings are of the slip-in, precision type which can be renewed after removing crankcase cover and connecting rod caps. Although not absolutely necessary, it is more convenient and time will be saved by removing the "Power-Trol" pump.

When installing new bearing shells, make certain that the bearing shell projections engage the milled slot in connecting rod and bearing cap and that cylinder numbers on rod and cap are in register and face toward top of engine. Number one cylinder is on left side of tractor. Connecting rod bearing inserts are available in standard as well as undersizes of 0.002, 0.004, 0.020 and 0.022 for series 520, 530, 620 and 630; 0.002, 0.004, 0.020, 0.022, 0.030 and 0.032 for series 720 and 730.

Check the crankshaft and bearing inserts against the values which follow:

Crankshaft Crankpin Diameter
    Series 520-530 . . . . . . .2.874 -2.875
    Series 620-630 . . . . . . .3.3736-3.3750
    Series 720-730 . . . . . . .3.748 -3.749

Rod Bearing Running Clearance
    Series 520-530 . . . . . . .0.001 -0.004
    Series 620-630 . . . . . . .0.0011-0.0041
    Series 720-730 . . . . . . .0.002 -0.005

Rod Bolt Tightening Torque
    Series 520-530-720-730. . 85 Ft.-Lbs.
    Series 620-630 . . . . .105 Ft.-Lbs.

CYLINDER BLOCK
All Models

54. **REMOVE AND REINSTALL.** To remove the cylinder block, first remove cylinder head as outlined in paragraph 30 or 31 and the connecting rod and piston units as outlined in paragraph 50. Remove spark plug covers, disconnect plug wires and remove spark plugs. Disconnect the heat indicator bulb and remove the upper water pipe rear casting (water outlet casting).

Remove cap screws (X—Fig. JD 1797) retaining cylinder block to front end support and remove stud nuts retaining block to main case. Slide block forward and withdraw from tractor.

Reinstall cylinder block by reversing the removal procedure and tighten the stud nuts which retain cylinder block to main case to the following torque:

    Series 520-530 . . . . .166 Ft.-Lbs.
    Series 620-630 . . . . .175 Ft.-Lbs.
    Series 720-730 . . . . .275 Ft.-Lbs.

CRANKSHAFT, SEALS AND MAIN BEARINGS

All Models

54A. The crankshaft is carried in two main bearing housings which are fitted with sleeve type bearings. The sleeve-type main bearings are available as individual repair parts and they must be sized after installation in the main bearing housings to provide the recommended shaft diametral clearance. Main bearings are also available already installed in the main bearing housings, on a factory exchange basis. The exchange main bearing and housing units are pre-sized and are available in standard size as well as undersizes of 0.002 and 0.004. Standard size main bearing bore is as follows:

520-530 (Right)	2.8785-2.8795
520-530 (Left)	2.2535-2.2545
620-630 (Right)	4.0050-4.0060
620-630 (Left)	2.7540-2.7550
720-730 (Right)	4.0050-4.0060
720-730 (Left)	3.2545-3.2555

Before removing the main bearing housings, check for excessive clearance between crankshaft and main bearings by mounting a dial indicator so that contact button is resting on crankshaft near the main bearing housings. Move crankshaft up and down in a manner similar to that shown in Fig. JD1798 and observe the main bearing clearance as shown on the dial indicator. Main bearing running clearance should be as follows:

520-530 (Right)	0.004 -0.006
520-530 (Left)	0.004 -0.006
620-630 (Right)	0.005 -0.007
620-630 (Left)	0.004 -0.006
720-730 (Right)	0.005 -0.007
720-730 (Left)	0.0045-0.0065

Fig. JD1798—Typical procedure for checking main bearing oil clearance. Tractor shown is a model 60, but procedure on series 520, 530, 620, 630, 720 & 730 is similar.

The right main bearing housing is fitted with a seal to prevent mixing of transmission oil with the engine crankcase oil. Oil leakage at the left main bearing is prevented by a packing and oil slinger (flywheel spacer).

55. **MAIN BEARINGS.** Although most repair jobs associated with the crankshaft and main bearings will require removal of both main bearing housings, there are infrequent instances where the failed or worn part is so located that repair can be accomplished safely by removing only one of the housings. In effecting such localized repairs, time will be saved by observing the following paragraphs as a general guide.

56. RIGHT MAIN BEARING HOUSING AND SEALS. To remove the right main bearing housing, first remove clutch and belt pulley as outlined in paragraph 118 for 520, 530, 620 and 630 or 119 for 720 and 730. Remove the reduction gear cover as in paragraphs 123, 130 or 136. Withdraw spacer or washer (S—Fig. JD1799) and remove the first reduction gear (RG). Remove the power shaft idler gear (IG), remove snap ring (SR) and using a puller as shown in Fig. JD1800, remove the power shaft drive gear but be careful not to damage end of crankshaft.

Fig. JD1799 — Right side of main case typical of that used on 620 & 630. Series 520, 530, 720 & 730 are similar.

DG. Power shaft drive gear
IG. Power shaft idler gear
RG. First reduction gear
S. Spacer or washer
SR. Snap ring

Fig. JD1800—Using puller to remove the continuous power shaft drive gear. Use care not to damage end of crankshaft during this operation.

Unbolt and withdraw the main bearing housing. NOTE: Housings have a slight press fit in main case and care should be exercised when removing them. Oil seal (24—Fig. JD1801) can be renewed at this time. Apply gun grease to lips of seal before installing housing.

Reinstall main bearing housing, using a thin sleeve (or shim stock) to guide oil seal over crankshaft and make certain that oil groove in main bearing housing and gaskets (13 and 20) are properly positioned over the oil feed hole in the main case.

Tighten the bearing housing retaining cap screws to a torque of 100-Ft.-Lbs. on 520 and 530, 150 Ft.-Lbs. on 620, 630, 720 and 730 and install safety wire in the drilled cap screw heads. Undrilled screws are secured by upsetting lock plates. Using a brass drift, install the powershaft drive gear as shown in Fig. JD1802.

57. LEFT MAIN BEARING HOUSING, OIL SLINGER & SEAL. To remove the left main bearing housing, first remove flywheel cover and flywheel as outlined in paragraph 59. Remove the oil slinger housing (3—Fig. JD1801). Mark the relative position of the oil slinger with respect to the crankshaft and remove the oil slinger (5). NOTE: The slinger is often called a flywheel spacer. Seal (6) which is

located inside the oil slinger can be renewed at this time.

Unbolt and withdraw the left main bearing housing and renew thrust washers (10) if they are damaged or show wear. NOTE: Housings have a slight press fit in main case and care should be exercised when removing them.

When reinstalling, make certain that oil groove in main bearing housing and gaskets (13 and 14) are properly positioned over the oil feed hole in the main case. Tighten the bearing housing retaining cap screws to a torque of 100 Ft.-Lbs. on 520 and 530, 150 Ft.-Lbs. on 620, 630, 720 and 730.

Check the inside diameter of the oil slinger housing at (D—Fig. JD1802A). Diameter should be 2.753-2.757 for 520 and 530 and 3.253-3.257 for 620, 630, 720 and 730. If the diameter is excessive, renew the housing. If diameter is too small, either enlarge the bore or install a new housing.

When reassembling, make certain that the previously affixed marks on crankshaft and oil slinger are aligned, install the oil slinger housing and tighten the housing retaining cap screws finger tight. Using feeler gage or shim stock (SS) as shown in Fig. JD1803, make certain that the oil slinger housing is perfectly centered

Fig. JD1802 — Using a hammer and brass drift to install the continuous power shaft drive gear.

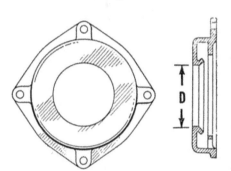

Fig. JD1802A — Check bore of left main bearing housing cover. The cover is sometimes called an oil slinger housing.

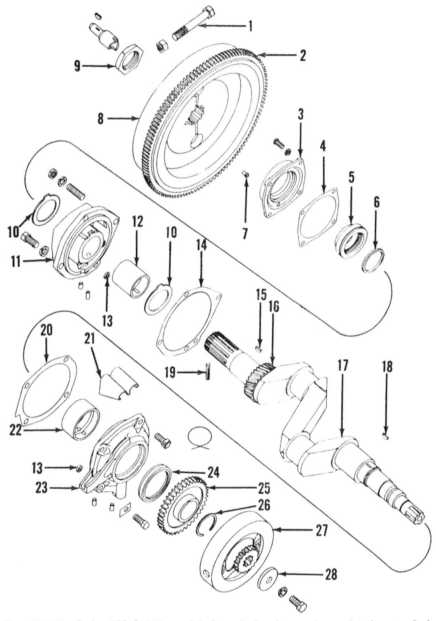

Fig. JD1801—Series 620 & 630 crankshaft, main bearings and associated parts. Series 520, 530, 720 and 730 are similarly constructed.

1. Flywheel clamp bolt	11. Left main bearing housing	20. Gasket
2. Flywheel ring gear		21. Transmission oil baffle
3. Cover	12. Left main bearing	22. Right main bearing
4. Gasket	13. Gasket	23. Right main bearing housing
5. Slinger (flywheel spacer)	14. Gasket	24. Oil seal
6. Packing ring	15. Woodruff key	25. Power shaft drive gear
7. Spacer drive pin	16. Crankshaft timing gear	26. Snap ring
8. Flywheel	17. Crankshaft	27. Clutch drive disc
9. Lock nut	18. Woodruff key	28. Washer
10. Thrust washer	19. Roll pin	

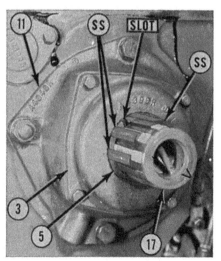

Fig. JD1803—Oil slinger housing (3) can be centered about the oil slinger (5) by inserting shim stock (SS) between oil slinger and housing before tightening the retaining cap screws. Notice the "V" mark on end of crankshaft. This mark must register with a similar mark on the flywheel.

about the slinger and tighten the cap screws.

NOTE: If position of oil slinger with respect to crankshaft is questionable, check the following: Observe right side of flywheel near the hub where a small drive pin is located. This pin must engage the slot in the oil slinger when "V" mark on left side of flywheel and "V" mark on end of crankshaft are in register.

When reassembling, adjust the crankshaft end play as outlined in paragraph 60.

58. **CRANKSHAFT.** To remove the crankshaft, remove "Powr-Trol" pump, crankcase cover, connecting rod caps and rod bearing inserts. Turn crankshaft and push the connecting rod and piston units forward. Support crankshaft and remove both main bearing housings as outlined in paragraphs 56 and 57. Withdraw crankshaft from main case being careful not to nick or damage the bearing journals.

Check the crankshaft and bearings against the values which follow:

Main Journal Diameter
520-530 (Right)	2.8735-2.8745
520-530 (Le`t)	2.2485-2.2495
620-630 (Right)	3.9990-4.0000
620-630 (Left)	2.7490-2.7500
720-730 (Right)	3.9990-4.0000
720-730 (Left)	3.2490-3.2500

Main Bearing Bushings Inside Diameter
520-530 (Right)	2.8785-2.8795
520-530 (Left)	2.2535-2.2545
620-630 (Right)	4.0050-4.0060
620-630 (Left)	2.7540-2.7550
720-730 (Right)	4.0050-4.0060
720-730 (Left)	3.2545-3.2555

Desired Main Bearing Clearance
520-530 (Right)	0.004 -0.006
520-530 (Left)	0.004 -0.006
620-630 (Right)	0.005 -0.007
620-630 (Left)	0.004 -0.006
720-730 (Right)	0.005 -0.007
720-730 (Left)	0.0045-0.0065

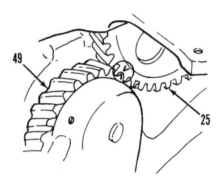

Fig. JD1804 — Valves are properly timed when "V" marks on camshaft gear and crankshaft gear are in register as shown.

Pulley Journal Diameter
520-530	2.1815-2.1825
620-630-720-730	2.432 -2.434

Pulley Bushing Inside Diameter (Minimum)
520-530	2.186
620-630-720-730	2.4375

Pulley Journal Clearance
520-530	0.0035-0.0045
620-630-720-730	0.0035-0.0055

Refer to paragraph 53 for crankpin diameter and rod bearing running clearances.

If crankshaft gear is damaged, it can be pulled from crankshaft at this time. Heat new gear in oil or water and install on crankshaft so that timing mark on gear is toward crankshaft web (inside of case) and shoulder on gear is toward end of crankshaft (outside of case).

When reassembling, make certain that "V" mark on crankshaft gear is in register with "V" mark on camshaft gear as shown in Fig. JD1804 and adjust the crankshaft end play as outlined in paragraph 60.

FLYWHEEL AND CRANKSHAFT END PLAY

All Models

59. To remove the flywheel, disconnect starter pedal linkage on series 620 or remove starter button on series 520 and 720. On all models remove flywheel cover and flywheel lock nut and loosen the flywheel clamp bolts. Attach hoist to flywheel in a suitable manner and bump or pull flywheel from crankshaft.

To install starter ring gear on flywheel, heat gear to approximately 550 degrees F. and place gear on flywheel so that beveled end of ring gear teeth face toward engine crankcase.

Fig. JD1805 — The recommended crankshaft end play of 0.005-0.010 can be checked by using a dial indicator as shown.

When installing flywheel, observe the engine side of the flywheel near the hub where a small drive pin is located. This pin must engage the slot in the crankshaft oil slinger when "V" mark on left side of flywheel and "V" mark on end of crankshaft are in register. Adjust the crankshaft end play as follows:

60. The crankshaft end play is controlled by the position of the flywheel on the crankshaft. To adjust the end play, proceed as follows: Drive flywheel on crankshaft and mount a dial indicator so that indicator contact button is resting on flywheel as shown in Fig. JD1805. Engage clutch, move clutch lever back and forth and observe crankshaft end play as shown on the dial indicator. Continue driving flywheel on crankshaft until the desired end play of 0.005-0.010 is obtained. Tighten both flywheel clamp bolts securely and install the lock nut. Secure lock nut in position by peening a portion of the nut into one of the crankshaft keyways.

CRANKCASE COVER

All Models

61. The crankcase cover can be removed for gasket renewal without removal of any other parts. If however, work is to be performed inside the crankcase, it is recommended that "Powr-Trol" pump also be removed.

Fig. JD1806 — Series 520 engine oil pressure can be adjusted by removing cap nut (CN) and turning the pressure adjusting screw as required. Series 530 is similar.

OIL PRESSURE

Series 520-530

62. Recommended oil pressure is 10-15 psi when engine is running at high idle speed. To check and/or adjust the pressure, proceed as follows: Disconnect the oil pressure gage line from connector in main case near rear of governor housing and connect the master gage to the connector. Start engine and observe oil pressure on master gage.

If pressure is not as specified, remove cap nut (CN—Fig. JD1806), loosen jam nut and turn adjusting screw **in** to increase pressure, **out** to decrease pressure.

Series 620-630-720-730

62A. Recommended oil pressure is 10-15 psi when engine is running at high idle speed. To check and/or adjust the pressure, proceed as follows: Disconnect the oil pressure gage line from connector at rear of governor case and connect the master gage to the connector. Start engine and observe oil pressure on master gage.

If pressure is not as specified, remove pipe plug (PP—Fig. JD1807) from right side of main case and turn pressure regulator screw **in** to increase pressure and **out** to decrease oil pressure.

FILTER HEAD AND REGULATOR

All Models

63. To remove the oil filter head, first drain the crankcase and remove the crankcase cover. Disconnect oil lines from filter head.

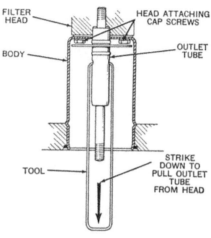

Fig. JD1808—Home made tool which can be used for removing the oil filter outlet as shown in Fig. JD1809. Tool can be made from 3/16-inch cold rolled steel rod.

Fig. JD1809—Using the tool shown in Fig. JD1808 to remove the oil filter outlet. Use a hammer to strike down on tool.

NOTE: If special tools are available, it is possible to disconnect these oil lines by working through the crankcase opening. Due to space limitations, it is often very time consuming, and if nipples connecting oil lines to filter head are not tight in filter head, it is oftentimes impossible for the average man to disconnect the oil lines by working through the crankcase opening; in which case, the following procedure is used.

63A. Disconnect connecting rods from crankshaft and remove carburetor, air inlet elbow, exhaust pipe and on 520, 530, 620 and 630, remove tool box. On all models, remove spark plug covers, disconnect spark plug wires and drain cooling system. Disconnect upper water pipe from cylinder block and remove lower water pipe. Remove valve cover, rocker arms assembly and push rods. Unbolt cylinder block from main case and front end support. Slide cylinder head and cylinder block assembly forward as far as possible. Working through front opening in main case, disconnect the oil lines from filter head.

63B. After oil lines are disconnected, remove the filter element. Using $\frac{3}{16}$-inch round, cold rolled steel rod, make up a tool similar to that shown in Fig. JD1808. Hook jaws of puller tool into holes of the filter outlet as shown in Fig. JD1809, then strike the

Fig. JD1807—Series 620 engine oil pressure can be adjusted by removing the pipe plug (PP) and turning the pressure adjusting screw as required. The pipe plug on series 630, 720 & 730 is similarly located.

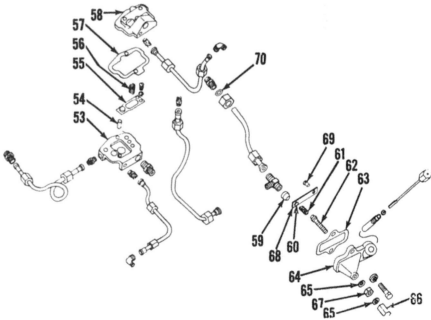

Fig. JD1810—Exploded view of series 520 oil filter head and pipes. Oil pressure is adjusted externally with screw (62). Series 530 is similar.

53. Filter head	57. Gasket	61. Coil spring	66. Cap nut
54. Dowel pin	58. Filter head cover	62. Adjusting screw	67. Jam nut
55. Leaf spring and button	59. Bushing	63. Gasket	68. Leaf spring
56. Coil spring	60. Relief spring pilot	64. Housing	69. Rivet
		65. Copper washer	70. Washer

bottom of the tool in a downward motion to remove the oil filter outlet. Working through bottom of filter body, remove the three cap screws retaining filter head to filter body.

Withdraw filter head and filter head cover through crankcase top opening. Make certain, however, that filter head assembly does not come apart and fall into crankcase.

Inspect filter head to make certain that mating and mounting surfaces of filter head are in good condition and not distorted. Check the condition of leaf springs and make sure that buttons in springs are tight. Oil pressure adjusting screws should turn freely. On 620, 630, 720 and 730, the adjusting screw lock spring should have sufficient tension to hold adjusting screw in position.

When reassembling, use light gage wire to hold filter head and filter head cover together while installing the cap screws retaining the filter head to the filter body. Using a suitable piece of pipe, drive the filter outlet tube into position. When installing oil filter element, tighten the cover nut only enough to eliminate oil leakage.

FILTER BODY

All Models

64. Zero or low oil pressure can be caused by a distorted oil filter body (87—Fig. JD1812). Distortion is usually caused by over tightening the filter bottom retaining nut. The nut (84) should be tightened only to eliminate oil leakage at this point. If leakage occurs with normal tightening, renew gasket (81).

To renew the filter body, first remove the filter head as outlined in paragraph 63 and proceed as follows: Place a cylindrical wooden block in filter body and jack up the block, thereby forcing the filter body out of the crankcase recess.

To install a new oil filter body, coat the portion below the "bead" with white lead or equivalent sealing compound to prevent leaks and facilitate drawing body into recess of main case. Temporarily place filter head on filter body to make certain that cap screw holes in body and head are aligned and oil lines will align with filter head connections; then lay filter head aside.

To draw filter body into crankcase recess, use a long bolt and two steel

plates as shown in Fig. JD1813. Tighten the nut until bead of body seats against crankcase. Install the remaining parts by reversing the removal procedure.

OIL PUMP

All Models

65. **BODY GEARS—RENEW.** To remove the body gears only, first drain crankcase and remove cover from pump body. Slide idler gear and shaft out of pump body as shown in Fig. JD1814. Remove crankcase cover and withdraw coupling (90—Fig. JD1815) from above. Pull drive gear and shaft from pump body.

Inspect the removed parts, using the values tabulated in paragraph 66A. Reassemble by reversing the disas-

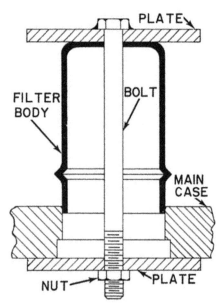

Fig. JD1813 — Suggested home-made tool for installing oil filter body. The long bolt can be welded to the upper plate.

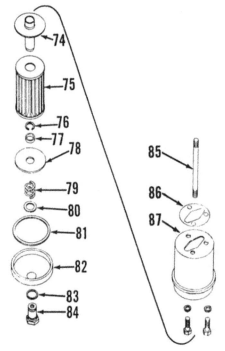

Fig. JD1811 — Typical exploded view of series 620, 630, 720 & 730 oil filter head and pipes. Oil pressure is adjusted with screw (62).

Fig. JD1812—Exploded view of oil filter. When renewing element (75), tighten nut (84) only enough to eliminate oil leakage.

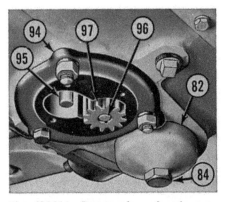

Fig. JD1814—Bottom view of main case, showing oil pump and filter installation. The oil pump cover and idler gear have been removed.

3. Pipe plug	59. Bushing
53. Filter head	60. Relief spring
54. Dowel pin	pilot
55. Leaf spring	61. Coil spring
56. Coil spring	71. Bracket
57. Gasket	72. Adjusting screw
58. Filter head cover	spring

74. Filter outlet	81. Gasket
76. Snap ring	82. Cover
77. Spacer	83. Copper gasket
78. End plate and	84. Nut
spacer	85. Stud
79. Spring	86. Gasket
80. Washer	87. Oil filter body

sembly procedure, making certain that coupling (90) engages the oil pump drive gear.

66. **R&R AND OVERHAUL.** To remove the complete oil pump, drain crankcase, remove crankcase cover and disconnect oil lines from oil pump body.

NOTE: If special tools are available, it is possible to disconnect these oil lines by working through the crankcase opening. Due to space limitations, it is often very time consuming, and if nipple connecting oil lines to pump are not tight in pump, it is oftentimes impossible for the average man to disconnect the oil lines by working through the crankcase opening; in which case, it will be necessary to move cylinder block forward and work through front opening in main case. Refer to paragraph 63A.

After the oil lines are disconnected, unbolt pump body from main case and withdraw pump assembly from below.

66A. Completely disassemble the pump and check the parts against the values which follow:

Drive shaft bore in
  pump body 0.627-0.628
Drive shaft diameter.........0.625
Drive shaft clearance in
  pump body 0.002-0.003
Diametral clearance be-
  tween gear teeth and
  pump body 0.002-0.006

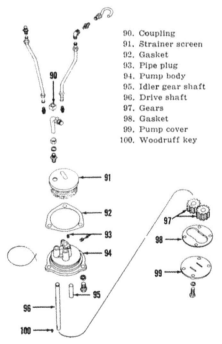

90. Coupling
91. Strainer screen
92. Gasket
93. Pipe plug
94. Pump body
95. Idler gear shaft
96. Drive shaft
97. Gears
98. Gasket
99. Pump cover
100. Woodruff key

Fig. JD1815—Exploded view of oil pump and associated pipes.

Gear bore in pump body .2.086-2.088
Gear diameter 2.082-2.084
Idler gear shaft
  diameter 0.6285-0.6290
Idler gear shaft bore
  in pump body........0.627-0.628
Clearance between
  gears and cover 0.008-0.021
Depth of body gear bore .1.240-1.245
Gear thickness 1.246-1.250
Thickness of cover gasket .0.018-0.022

When reinstalling oil pump, make certain that coupling (90) engages the oil pump drive gear.

CARBURETOR
(Not LP-Gas)

Series 520-530-620-630-720-730

68. Marvel-Schebler carburetors are used and their applications are as follows:

520 & 530 Gasoline......DLTX- 99
520 & 530 All-Fuel.....DLTX- 96
620 Orchard Gasoline
  Prior 6223000 ....DLTX- 94
620 Orchard Gasoline
  After 6222999 ....DLTX-106
Other Series 620
  Gasoline ....DLTX- 94
630 Gasoline DLTX-106
620 & 630 All-Fuel.....DLTX- 97
720 & 730 Gasoline.....DLTX- 95
720 & 730 All-Fuel.....DLTX- 98

Refer to Appendix I in this manual for carburetor calibration data.

LP-GAS SYSTEM

Total fuel tank capacity is 39 gallons on 620, 630, 720 and 730; 28 gallons on 520 and 530. However, tank should NEVER be filled more than 85 per cent full of fuel (33 gallons on 620, 630, 720 and 730; 24 gallons on 520 and 530). This allows room for expansion of the fuel due to a possible rise in temperature.

CAUTION: LP-Gas expands readily with any decided increase in temperature. If tractor must be taken into a warm shop to be worked on during extremely cold weather, make certain that fuel tank is as near empty as possible. LP-Gas tractors should never be stored or worked on in an unventilated space

TROUBLE-SHOOTING

All Models

70. The following trouble shooting paragraphs list troubles that can be attributed directly to the fuel system; however, many of the troubles can be caused by derangement of other parts such as valves, battery, spark plugs, distributor, coil, resistor, etc.

The procedure for remedying many of the causes of trouble is evident. The following paragraphs will list the most likely causes of trouble, but only the remedies which are not evident.

71. **HARD STARTING.** Hard starting could be caused by:
  a. Improperly blended fuel.
  b. Excess-flow valve in withdrawal valve closed. Close withdrawal valve to reset excess-flow valve, then open withdrawal valve slowly.
  c. Incorrect starting procedure.
  d. Restricted fuel strainer.
  e. Liquid fuel in lines.
  f. Automatic fuel shut-off on strainer not operating properly. A "click" should be heard when ignition switch is turned on. If no click is heard, check wiring and check solenoid on strainer.
  g. Plugged vent on back of convertor. The vent is a ¼-inch tapped hole.
  h. Defective low pressure diaphragm in convertor.
  i. Stuck high pressure valve or broken high pressure spring in convertor.
  j. Restricted fuel lines.

72. **ENGINE SHOWS NOTICEABLE LOSS OF POWER:**
  a. Throttle not opened sufficiently due to maladjusted governor or carburetor linkage.
  b. Plugged vent on back of convertor. The vent is a ¼-inch tapped hole.
  c. Clogged fuel strainer (if strainer shows frost, it is probably clogged).
  d. Plugged fuel lines or restrictions in withdrawal valves (indicated by frost). With engine cold, both withdrawal valves closed and lines and filter empty of gas, remove plug at bottom of strainer, open the liquid withdrawal valve slightly and check for fuel flow.
  e. Closed excess flow valves in vapor or liquid withdrawal valves (Indicated by frosted withdrawal valve). Close frosted valve to seat excess flow valve, then re-open slowly.

f. Lean mixture caused by restricted or altered fuel lines or hoses.

g. Sticking high pressure valve in convertor.

h. Restricted low pressure valve in convertor.

i. Defective convertor diaphragms.

j. Engine not up to operating temperature. Check thermostat.

k. Improperly adjusted carburetor.

l. Faulty adjustment of throttle linkage.

m. Faulty carburetor adjustment.

n. Faulty gasket between carburetor and manifold.

o. Leaking fuel hose between convertor and carburetor.

p. Air entering between carburetor throttle body and air horn.

q. Clogged air filter.

73. POOR FUEL ECONOMY. Could be caused by any of the conditions listed in paragraph 72, plus:

a. Improperly filled fuel tank.

b. Faulty fuel.

74. ROUGH IDLING. Could be caused by faulty ignition system plus:

a. Faulty adjustment of carburetor.

b. Faulty adjustment of throttle linkage.

c. Faulty carburetor to manifold gasket.

d. Leaking hose between carburetor and convertor.

75. POOR ACCELERATION.

a. Faulty idle speed adjustment.

b. Faulty load adjustment.

c. Faulty low pressure diaphragm in convertor.

d. Restricted convertor to carburetor hose.

76. ENGINE STOPS WHEN THROTTLE IS BROUGHT TO SLOW IDLE POSITION.

a. Faulty slow idle speed adjustment.

b. Faulty convertor to carburetor hose.

c. Faulty carburetor gaskets.

d. Faulty gasket between carburetor and air horn.

e. Leaking convertor back cover gasket (Indicated by fuel bubbles in radiator).

77. OVERHEATING. Could be caused by defective cooling system plus:

a. Lean mixture due to faulty adjustment of linkage.

78. CONVERTOR FREEZES UP WHEN ENGINE IS COLD.

a. Running on liquid fuel before engine is warm.

b. Leaking convertor high pressure valve. With ignition switch turned on, this can be detected by odor of gas or hissing sound.

79. CONVERTOR FREEZES UP DURING NORMAL OPERATION. Could be caused by a defective cooling system plus:

a. Water circulating backwards through convertor.

b. Restrictions in water piping or convertor.

c. Running on liquid fuel before engine is warmed up.

80. FROST ON WITHDRAWAL VALVE.

a. Closed excess flow valve. Close withdrawal valve to reset the excess flow valve; then, open withdrawal valve slowly.

b. Water in fuel tank will sometimes cause ice to form in liquid withdrawal valve. Empty all fuel from filler hose, pour one pint of alcohol in hose, attach hose to fuel tank and inject alcohol into fuel tank. Alcohol will act as an antifreeze and water will be dissipated through the engine.

81. LACK OF FUEL AT CARBURETOR.

a. Empty fuel tank or withdrawal valve closed.

b. Excess flow valve in withdrawal valve closed. Close the withdrawal valve to reset the excess flow valve; then, open the withdrawal valve slowly.

c. Restriction in withdrawal valve. See paragraph 80.

d. Restricted fuel strainer.

e. Faulty wiring to strainer shut off valve or faulty valve.

f. Faulty convertor high pressure valve.

g. Restricted fuel lines.

82. FUEL IN COOLING SYSTEM. This trouble is usually caused by a ruptured convertor back cover gasket.

CARBURETOR

All Models

87. ADJUSTMENT. Two speed adjustments, two idle mixture adjusting needles and one load adjustment screw is provided on the carburetor and linkage.

87A. LINKAGE ADJUSTMENTS. With throttle rod (Fig. JD1816) assembled and carburetor throttle lever and speed control lever in wide open position, adjust the length of the rod ½-hole short as follows: Turn the swivel lock nuts (Fig. JD1817) until

Fig. JD1818—520, 530, 620 and 630 LP-Gas carburetor idle mixture adjusting needles. Series 720 & 730 are similar.

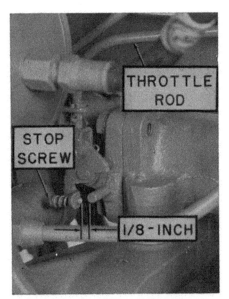

Fig. JD1816 — Throttle rod and throttle stop screw installation on LP-Gas carburetors.

Fig. JD1817—520 & 530 governor and carburetor linkage used on LP-Gas tractors. The parts used on series 620, 630, 720 & 730 are similarly located.

they clear the swivel; then, position the forward lock nut ⅛-inch from swivel and tighten the rear nut.

With speed control lever in closed position, hold throttle lever closed and adjust the throttle stop screw so the engine runs at 300 rpm.

Adjust speed control rod at ball joint until engine idles at 600 rpm, but make certain that governor spring clears governor arm stop screw during this adjustment. Now, with engine stopped and speed control lever in closed position, adjust the governor arm stop screw so that the throttle stop screw (Fig. JD1816) clears the stop pin by ⅛-inch.

87B. IDLE MIXTURE. With speed control lever in slow idle position,

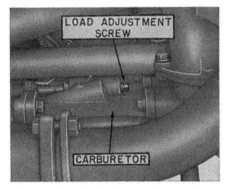

Fig. JD1819—LP-Gas carburetor load adjusting screw.

hook a spring or heavy rubber band over throttle lever to hold throttle closed on stop screw; then adjust the stop screw to obtain an engine speed of 600 rpm. Adjust each of the idle needles (Fig. JD1818) individually to obtain maximum engine rpm. Operate engine from slow idle to fast idle several times. If acceleration and deceleration is not smooth and even, open the idle needles slightly. Reset throttle stop screw to obtain a slow idle speed of 300 rpm and remove the rubber band or spring. Recheck the linkage adjustments as outlined in paragraph 87A.

87C. FAST IDLE ADJUSTMENT. With the speed control lever all the way forward, adjust the cap screw (S—Fig. JD1817) to obtain an engine high idle speed of 1460 rpm for 520 and 530, 1260 rpm for 620, 630, 720 and 730.

87D. LOAD ADJUSTMENT. Load adjustment screw (Fig. JD1819) on carburetor should be adjusted to provide the best operation under loaded conditions. Average adjustment is three turns open for 520 and 530, 2¼-2½ turns open for 620, 630, 720 and 730.

FUEL STRAINER AND SHUT-OFF VALVE

All Models

88. All of the fuel must pass through the strainer before reaching the convertor. The strainer contains a filter element which consists of a felt pad and a chamois disc backed by a brass screen on each side. The purpose of the filter is to remove all solids from the fuel before the fuel reaches the convertor valves. A solenoid operated, automatic fuel shut-off is located on top of the strainer. Whenever the ignition switch is turned on, the solenoid opens the valve with an audible "click."

If the strainer shows frost, it is probably clogged and needs cleaning.

89. **CLEANING.** To clean the strainer first make certain that both fuel tank withdrawal valves are closed, engine is cold and lines and filter are empty of gas. Note: Lines and filter will be empty if engine was properly stopped.

Remove plug (39—Fig. JD1820) from bottom of strainer and open the liquid withdrawal valve slightly; thus allowing pressure from the fuel tank to blow out any accumulation of dirt.

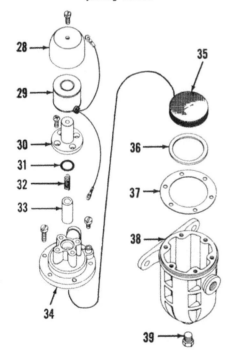

Fig. JD1820—Exploded view of LP-Gas fuel strainer and automatic fuel shut-off.

28. Case	34. Strainer cover
29. Solenoid coil	35. Filter pack
30. Plunger housing	36. Retainer ring
31. "O" ring	37. Gasket
32. Spring	38. Strainer body
33. Plunger	39. Drain plug

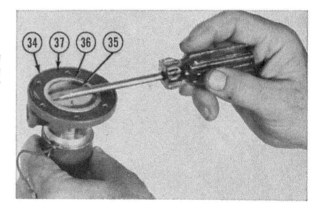

Fig. JD1821 — Removing filter pack retaining ring from LP-Gas fuel strainer cover.

34. Strainer cover
35. Filter pack
36. Retainer ring
37. Gasket

Fig. JD1822 — Removing the fuel shut-off plunger and plunger housing from LP-Gas fuel strainer cover.

28. Case
29. Solenoid coil
30. Plunger housing
33. Plunger
34. Strainer cover

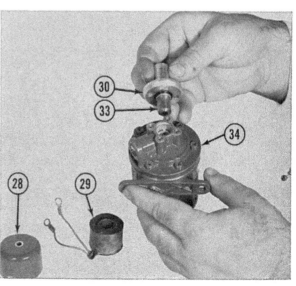

90. R&R AND OVERHAUL. To remove the strainer, close both fuel tank withdrawal valves, disconnect fuel lines and remove strainer.

Remove cover from strainer body and remove the filter pack by prying out the retainer ring with a screw driver as shown in Fig. JD1821. Filter pack can be cleaned in a suitable solvent. Reinstall filter pack with chamois disc toward top.

Disconnect wire and remove case from shut-off valve. Lift off coil (29—Fig. JD1822), remove plunger housing (30) and lift out plunger and spring. Inspect and renew any damaged parts.

Assemble the solenoid coil, plunger and spring on plunger housing.

To test solenoid, connect it to a battery to see if plunger compresses spring in plunger housing.

When reassembling, fasten one of the coil wires to screw which retains case (28). After unit is installed on tractor, test for leaks by using soapy water around all connections.

CONVERTOR
Series 520-620-720

92. OVERHAUL. The high pressure valve can be removed and overhauled without removing convertor from tractor; any other work, however, cannot be accomplished until convertor has been removed.

93. HIGH PRESSURE VALVE. To remove the high pressure valve for cleaning and/or parts renewal, close both tank withdrawal valves, disconnect strainer-to-convertor fuel line and unscrew the high pressure valve jet from the convertor body.

Clean the valve assembly in a suitable solvent and renew any damaged parts. When reassembling, large end of spring (66—Fig. JD1823) goes toward aluminum seat (64). After valve is reinstalled, test for leaks with soapy water.

94. LOW PRESSURE DIAPHRAGM AND VALVE. With convertor removed from tractor, remove the convertor front cover (75—Fig. JD1823), low pressure diaphragm (72) and backing plate (69). Remove cotter pin from one end of primary lever pin (60) and remove the lever pin, primary lever (54), spring (62), low pressure valve seat (53) and low pressure valve (51). Inspect all parts thoroughly and renew any which are excessively worn.

When renewing the neoprene valve seat (53—Fig. JD1823), hinge it loosely to the primary lever so the seat can pivot on pin (52) and seat itself properly on the valve. When reassembling, hook end of primary lever under the secondary lever and using a straight edge and rule, measure distance from end of secondary lever to edge of body casting as shown in Fig. JD1826. Bend the primary lever at (P), if necessary, until the measured distance is $\frac{1}{8}$-$\frac{5}{32}$-inch as shown.

94A. Install diaphragm backing plate with "dish" down toward body and low pressure diaphragm with "dish" up toward cover. Bottom of front cover goes toward convertor attaching lug which is partially ground off. NOTE: Before completely tightening the cover cap screws, make certain that low pressure diaphragm has sufficient sag to operate properly. Also, make certain that there are no wrinkles along edge of diaphragm.

95. HIGH PRESSURE DIAPHRAGM. With convertor removed from tractor, remove the convertor front cover (75—Fig. JD1823), low pressure diaphragm (72) and backing plate (69). Remove the high pressure diaphragm cover and lift off the diaphragm with spring and link.

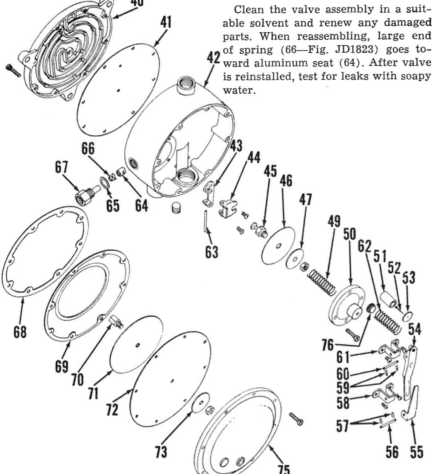

Fig. JD1823—Exploded view of 520, 620 & 720 LP-Gas convertor.

40. Back cover	50. Cover	57. Hair pin	65. Gasket
41. Gasket	51. Low pressure valve	58. Low pressure valve lever bracket	66. Spring
42. Convertor body	52. Pin		67. High pressure jet
43. High pressure valve operating bracket	53. Low pressure valve seat	59. Hair pin	68. Gasket
44. Lever	54. Low pressure valve primary lever	60. Pivot pin	69. Backing plate
45. Diaphragm link		61. Primary lever bracket	70. Diaphragm button
46. High pressure diaphragm	55. Secondary lever	62. Spring	71. Diaphragm plate
47. Diaphragm plate	56. Pivot pin	63. Pivot pin	72. Low pressure diaphragm
49. Spring		64. High pressure valve	73. Diaphragm plate
			75. Front cover

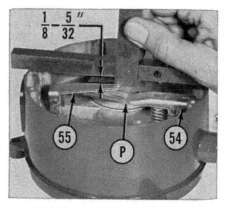

Fig. JD1826—Bend 520, 620 & 720 primary lever (54) at point (P) so that distance from end of secondary lever (55) to edge of convertor body is 1/8-5/32-inch as shown.

When reassembling, hook button on diaphragm link (45) under operating lever (44) and install spring and valve cover. Make certain that there are no wrinkles along edge of diaphragm before tightening the cover screws. Install low pressure diaphragm and front cover as in paragraph 94A.

96. HEAT EXCHANGER. If convertor shows frost when engine is warm, and if the pipes carrying water to and from the convertor are not plugged, remove convertor from tractor and take off the back cover. Thoroughly clean the water chamber and renew the back cover gasket. Bleed hole in gasket goes toward top of convertor and milled-off attaching lug on cover goes toward bottom.

Series 530-630-730

97. OVERHAUL. The high pressure valve seat can be removed for cleaning without removing convertor from tractor; any other work, however, cannot be accomplished until convertor has been removed. Before disassembling convertor, index mark the body and the front and back covers for reassembly reference.

97A. HIGH PRESSURE VALVE SEAT. The high pressure valve seat (7—Fig. JD1827) can be removed and cleaned in a suitable solvent as follows: Close both fuel tank withdrawal valves, disconnect the strainer-to-convertor fuel line and remove the inlet cover (5) and gasket (6). When reinstallation is complete, test for leaks with soapy water.

97B. LOW PRESSURE DIAPHRAGM AND VALVE. With convertor removed from tractor, remove front cover (36), low pressure diaphragm (33) and plates (32 and 34). Extract pin (16) and remove lever (15), spring (12) and seat (14). Renew any worn or damaged parts.

When reassembling, use a straight edge and rule as shown in Fig. JD 1827A and measure distance from end of loading lever to edge of body casting. Bend lever to obtain a distance of $\frac{5}{16}$-inch as shown.

Before tightening the cover cap screws, make certain that the low pressure diaphragm has sufficient sag to operate properly. Also, make certain that there are no wrinkles along edge of diaphragm.

97C. HIGH PRESSURE DIAPHRAGM. With convertor removed from tractor, remove front cover and low pressure diaphragm assembly. Remove the high pressure diaphragm cover (19—Fig. JD1827) and lift off the diaphragm assembly.

Fig. JD1827 — Exploded view of 530, 630 and 730 LP-Gas convertor.

1. Screw
2. Back cover
3. Gasket
4. Screw
5. Inlet cover
6. Gasket
7. High pressure valve seat
8. Lock spring
9. High pressure valve lever
10. Pin
11. Body
12. Low pressure spring
13. Valve seat pin
14. Low pressure valve seat
15. Low pressure valve lever
16. Pin
17. Screw
18. Screw
19. High pressure diaphragm cover
20. High pressure spring
21. Screw
22. Diaphragm plate
23. High pressure diaphragm
24. Diaphragm damper
25. Diaphragm washer
26. Diaphragm link
28. Gasket
29. Drain plug
30. Gasket
31. Diaphragm button
32. Diaphragm plate, large
33. Low pressure diaphragm
34. Diaphragm plate, small
35. Screw
36. Front cover
37. Screw

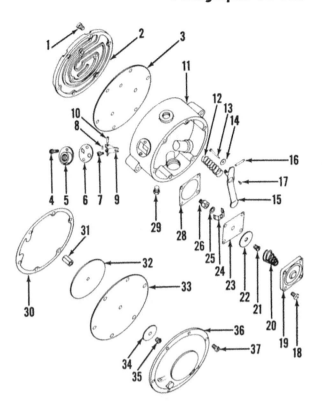

When reassembling, be sure button on link (26) is hooked under lever (9) and install spring and valve cover. Make certain that there are no wrinkles along edge of diaphragm before tightening the cover screws. Install low pressure diaphragm as in paragraph 97B.

97D. HEAT EXCHANGER. If convertor shows frost when engine is warm, and if pipes carrying water to and from the convertor are not plugged, remove convertor from tractor, take off back cover (2—Fig. JD1827) and thoroughly clean water chamber.

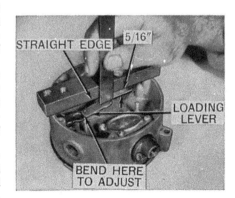

Fig. JD1827A—On 530, 630 and 730, bend loading lever as shown to obtain the 5/16-inch dimension.

GOVERNOR

All models are equipped with a centrifugal flyweight type governor which is driven by the engine camshaft gear. An idler gear, which is mounted in the rear portion of the governor case drives the live "Powr-Trol" pump. The fan drive pinion is mounted on the governor shaft, and is in constant mesh with the fan drive bevel gear.

SPEED AND LINKAGE ADJUSTMENT

All Models (Except LP-Gas)

100. LINKAGE ADJUSTMENTS. Before attempting to adjust the engine speed, free-up and align all linkage to remove any binding tendency and adjust or renew any parts causing lost motion. Cap screws (15—Fig. JD1828) should be adjusted so that a pull of 10-15 pounds at end of speed control lever is required to move lever through full range of travel when rod (7) is disconnected from speed control arm (12).

With the throttle rod (Fig. JD1829) assembled, and carburetor throttle lever and speed control lever in wide open position, adjust the length of the rod ½-hole short as follows: Turn the swivel lock nuts (N) until they clear the swivel; then, position the forward lock nut ⅛-inch from swivel and tighten the rear nut.

With the speed control lever in closed position, hold throttle lever closed and adjust the throttle stop screw (Fig. JD1830) so the engine runs at 300 rpm.

Adjust speed control rod (Fig. JD-1829) at ball joint until engine idles at 600 rpm, but make certain that governor spring clears governor arm stop screw during this adjustment.

Fig. JD1830—620 & 630 carburetor installation showing the throttle stop screw and mixture adjustment screws. Series 520, 530, 720 and 730 are similar.

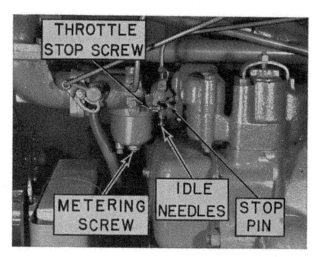

Fig. JD1828 — Exploded view of series 520 governor controls and throttle linkage. The general location of parts on series 530, 620, 630, 720 and 730 is similar.

1. Throttle rod
2. Yoke spring
3. Throttle rod yoke
4. Throttle rod swivel
5. Governor arm
6. Stop screw
7. Speed control rod
8. Speed control lever
9. Dowel pin, 2 used
10. Speed control plate with facings
11. Speed control rod ball joint
12. Speed control arm
13. Spacer
14. Spring
15. Cap screw
16. Governor spring

Fig. JD1829—Series 520 governor and throttle control linkage. Engine high idle speed is adjusted with screw (S). The linkage on series 530, 620, 630, 720 and 730 is similar.

Now, with engine stopped and speed control lever in closed position, adjust the governor arm stop screw (Fig. JD1829) so that the throttle stop screw (Fig. JD1830) clears the stop pin by ⅛-inch.

100A. IDLE MIXTURE. With speed control lever in slow idle position, hook a spring or heavy rubber band over throttle lever to hold throttle closed on stop screw; then adjust the stop screw to obtain an engine speed of 600 rpm. Short out No. 1 (left) spark plug and with engine running on the No. 2 cylinder, adjust the right hand idle adjusting needle (Fig. JD-1830) to obtain the maximum speed and record the rpm reading. Short out No. 2 (right) spark plug and with engine running on the No. 1 cylinder, adjust the left hand idle adjusting needle to obtain the maximum speed and record the rpm reading. Now, turn the idle needle of the cylinder having the higher rpm until it is the same as the speed of the slower cylinder. Reset throttle stop screw to obtain a slow idle speed of 300 rpm and remove the rubber band or spring. Recheck the linkage adjustments as outlined in paragraph 100.

100B. FAST IDLE ADJUSTMENT. With the speed control lever all the way forward, adjust the cap screw (S—Fig. JD1829) to obtain an engine high idle speed of 1460 rpm for 520 and 530, 1260 rpm for 620, 630, 720 and 730.

100C. METERING SCREW ADJUSTMENT. Metering screw (Fig. JD-1830) on carburetor should be adjusted to provide the best operating conditions. Average adjustment is 2¼ turns open.

All Models (LP-Gas)

100D. Speed and linkage adjustments for LP-Gas burning tractors are covered in paragraphs 87-87D.

OVERHAUL

All Models

101. Normal overhaul of the governor consists of removing and overhauling the shaft assembly only; and can be accomplished without removing governor housing from tractor. If, however, the fan drive bevel gear is damaged it will be necessary to remove the complete governor assembly as well as the fan shaft assembly in order to renew the matched set of bevel gears. Refer to paragraph 102 or 103 for removal of governor.

101A. **SHAFT AND WEIGHTS.** To overhaul the governor shaft and weights, first loosen set screw (SS—Fig. JD1828) and bump governor arm (5) up and off the governor lever shaft. Remove bearing housing (17—Figs. JD1831 and JD1832) and save shims (18) for reinstallation. Turn flywheel until governor weights are on top and bottom (vertical position) and withdraw governor shaft assembly as shown in Fig. JD1833.

Inspect all parts for damage or excessive wear. Inspect the bearing on right end of shaft for evidence of turning in governor case. If bearing has been turning in governor case, check the case by installing a new bearing cup. If a new cup fits loosely, it will be necessary to renew the governor case or reclaim same as in paragraph 104. If the fan drive bevel pinion is damaged, press governor

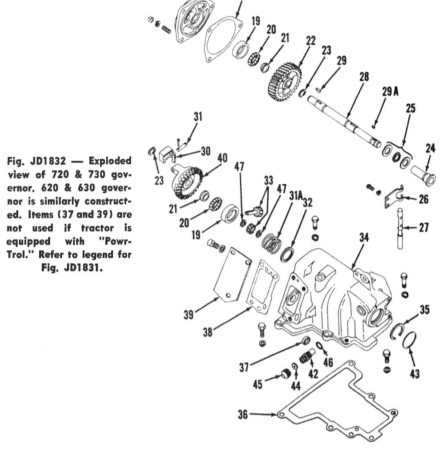

Fig. JD1832 — Exploded view of 720 & 730 governor. 620 & 630 governor is similarly constructed. Items (37 and 39) are not used if tractor is equipped with "Powr-Trol." Refer to legend for Fig. JD1831.

Fig. JD1831 — Exploded view of 520 & 530 governor assembly. Backlash between the fan drive bevel gears is controlled by shims (18). Items (37 and and 39) are not used if tractor is equipped with "Powr-Trol."

17. Left bearing housing
18. Shim gaskets
19. Bearing cup
20. Balls and retainer
21. Bearing cone
22. Drive gear
23. Snap ring
24. Sleeve
25. Thrust bearing
26. Governor lever
27. Lever shaft
28. Governor shaft
29. Woodruff key
29A. Woodruff key
30. Flyweight
31. Pin
31A. Bearing spring
32. Washer
33. Speed-hour meter drive and driven gears
34. Housing
35. Snap ring
36. Gasket
37. Plug (no "Powr-Trol")
38. Gasket
39. Cover
40. Fan drive bevel pinion
41. Snap ring
42. Cable connector
43. Expansion plug
44. Gasket
45. Cap
46. Washer
47. Snap ring

shaft out of pinion and renew the bevel gears which are available in a matched set only.

NOTE: If necessary to renew the bevel gears, it will also be necessary to remove the fanshaft assembly and adjust the mesh and backlash of the bevel gears as outlined in paragraph 108.

The spur drive gear (22) can be pulled or pressed from governor shaft if renewal is required. Thrust bearing (25) should be in good condition and sleeve (24) should slide freely on the governor shaft.

The governor lever shaft (27) should have a clearance of 0.002-0.004 in governor case. Bearing spring (31A) should require 99-121 pounds to compress it to a height of $1\frac{1}{16}$ inches.

When reassembling, use Figs. JD-1831 and JD1832 as a guide. Press the fan drive bevel pinion on governor shaft (if removed) until shoulder on pinion contacts snap ring on shaft. Press new drive gear on shaft until shoulder on gear contacts snap ring on shaft. On series 620 and 720, long hub of drive gear goes toward left end of shaft.

101B. Install governor shaft assembly by reversing the removal procedure and if fan drive bevel pinion was not renewed, install same number and thickness of shims (18—Figs. JD1831 and JD1832) as were originally removed so as to retain the desired bevel gear backlash of 0.004-0.007 for series 520, 530, 620 and 630; 0.004-0.006 for series 720 and 730.

REMOVE AND REINSTALL

As previously outlined in paragraph 101, the governor housing need not be removed to facilitate a normal overhaul of the governor shaft and weights. The subsequent paragraphs, however, will outline the procedure for removing the governor housing assembly to perform any of the following jobs.

A. Removal of engine camshaft and associated parts.

B. Removal of fan shaft assembly.

C. Renewal of fan drive bevel gears.

D. Reclaiming of governor case.

Series 520-530

102. To remove governor housing assembly, remove flywheel cover and flywheel. Drain the hydraulic "Powr-Trol" system and remove hydraulic pump. Disconnect speed control rod from governor spring and throttle rod from governor arm. Unbolt the fanshaft rear bearing housing from governor housing and governor housing from main case, noting the position of the dowel type cap screws. Disconnect oil pressure line from bushing in main case, located at right rear of governor housing. Raise governor until governor gear and camshaft gear are out of mesh, move governor rearward until the fan drive bevel gears

Fig. JD1833—Removing 520 & 530 governor shaft. Notice that weights are in vertical position. Other models are similar except flyweights are mounted on the fan drive bevel pinion.

Fig. JD1834 — Series 520 governor and "Powr-Trol" pump installation. Governor weight unit can be overhauled without removing governor housing from tractor. Series 530 is similar.

are out of mesh and withdraw governor assembly from tractor.

NOTE: Be careful not to lose the fan drive bevel gear mesh adjusting shims which are located between the fanshaft rear bearing housing and the governor case. If the bevel gears are not to be renewed, the same shims should be used during reassembly.

Install governor by reversing the removal procedure and adjust crankshaft end play as outlined in paragraph 59. The 520 governor installation is shown in Fig. JD1834.

Series 620-630-720-730

103. To remove the governor housing assembly, remove grille and loosen the two cap screws retaining front of hood to the radiator top tank. Remove two cap screws retaining the steering shaft and hood support to the gear shift quadrant. Disconnect oil lines from the automatic fuel shut-off valve located at top of fuel filter and disconnect oil line from connector located at right rear of governor case. Disconnect speed control rod from governor spring, throttle rod from governor arm, choke rod and fuel line from carburetor, coil wire from coil and fuel tank support from governor housing. On 720 and 730 it is also necessary to unbolt the fuel tank support from the upper water pipe rear casting. On all models, raise hood and fuel tank assembly and block-up between the steering shaft and hood support and the gear shift quadrant.

Drain the hydraulic "Powr-Trol" system and remove the hydraulic pump. Unscrew packing gland which retains the ventilator pipe to the vent pump cover and unbolt the fanshaft rear bearing housing from the governor housing. Unbolt front of fanshaft from its support.

Unbolt governor housing from main case, raise governor until governor gear and camshaft gear are out of mesh, move governor rearward until the fan drive bevel gears are out of mesh and withdraw governor assembly from tractor.

NOTE: Be careful not to lose the fan drive bevel gear mesh adjusting shims which are located between the fanshaft rear bearing housing and the governor case. If the bevel gears are not to be renewed, the same shims should be used during reassembly.

Install governor by reversing the removal procedure.

RECLAIMING GOVERNOR CASE

104. If the governor shaft right hand bearing cup has been turning in governor housing, and if a new bearing cup fits loosely in the housing bore, it will be necessary to renew housing, or reclaim same. To reclaim housing,

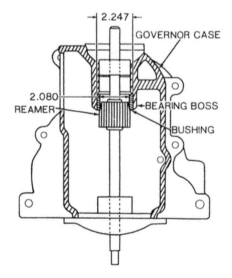

Fig. JD1835—Using a Woodring and Wise reclaiming fixture on governor case. See text. The fixture is available from Woodring and Wise Machine Co., Waterloo, Iowa.

remove and disassemble governor and proceed as follows:

Using a Woodring and Wise reclaiming fixture as shown in Fig. JD1835, ream the governor housing bearing bore to an inside diameter of 2.247 and to a depth at which the back side of the reamer is flush with inner edge of bearing boss. Obtain a John Deere F152R bushing and cut the bushing to a width of ⅝-inch. Press bushing in governor housing and ream the bushing to an inside diameter of 2.080. Reassemble governor by reversing the disassembly procedure.

COOLING SYSTEM
RADIATOR
Series 520-620

105. To remove radiator, it is first necessary to remove the hood and fuel tank assembly as follows: Drain cooling system and remove grille, steering wheel and Woodruff key. Remove the steering wormshaft front bearing housing and turn the wormshaft forward and out of housing; OR, unbolt worm housing from pedestal and remove wormshaft and housing assembly from tractor. Disconnect choke rod at carburetor, remove fuel line and disconnect oil lines from the automatic fuel shut-off valve which is located on top of fuel strainer. Remove muffler. Disconnect battery cable, remove the four cap screws retaining instrument panel to support, disconnect oil line from oil pressure gage and pull instrument panel rearward. Disconnect coil and regulator wires at instrument panel. Unscrew pull knob from choke rod and pull choke rod forward until free from rear support. Unclip heat indicator bulb wire from hood. Unbolt hood and lift hood and fuel tank assembly from tractor.

Remove air intake and exhaust pipes and loosen the air cleaner hose clamps. The air cleaner can be removed for convenience. Unbolt baffle plates from the radiator lower tank. Loosen fan belt and unbolt water pump from lower tank. Remove the cap screws retaining the steering pedestal to tractor and turn the pedestal, as required, to provide clearance for removing the radiator. Unbolt radiator from front end support and using a hoist, lift radiator assembly from tractor.

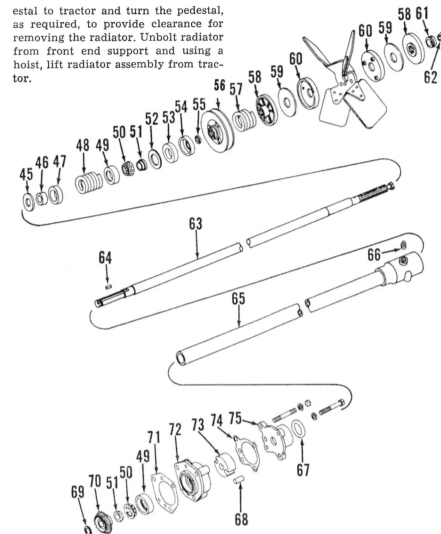

Fig. JD1837—Exploded view of 520, 530, 620 & 630 fan shaft and crankcase ventilator pump as used on models with manual steering. Shim gaskets (71) control mesh position of the fan drive bevel gears. Series 720 & 730 are similar except there is a washer located between items (47 and 48). On models with power steering, refer also to Fig. JD1766.

45. Washer	54. Felt retainer	62. Half-moon locks	69. Snap ring
46. Packing	55. "O" ring	63. Fan shaft	70. Bevel gear
47. Retainer	56. Pulley	64. Key	71. Shim gasket
48. Spring	57. Spring	65. Front bearing housing and tube	72. Fan shaft rear bearing housing (pump housing)
49. Bearing cup	58. Friction disc	66. Pipe plug	73. Pump rotor
50. Balls & retainer	59. Friction washer	67. "O" ring	74. Gasket
51. Bearing cone	60. Fan drive cup	68. Vent pump roller	75. Vent pump cover
52. Washer	61. Fan keeper		
53. Felt washer			

Series 530-630-730

105A. To remove radiator, drain cooling system and remove grille and hood. Unbolt and remove wormshaft and housing assembly from pedestal. Remove air cleaner and exhaust pipe. Unbolt baffle plates from the radiator lower tank. Loosen fan belt and unbolt water pump from lower tank. Remove the cap screws retaining the steering pedestal to the tractor, turn the pedestal as required to provide clearance, then unbolt and remove radiator assembly from tractor.

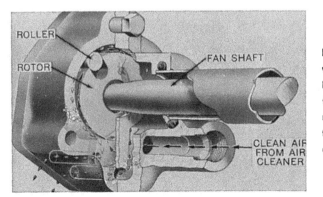

Fig. JD1836 — Cut-away view of crankcase ventilator pump. The pump rotor is mounted on the rear of the fan shaft and the pump housing is bolted to the governor case.

Series 720

106. To remove radiator, drain cooling system and remove grille, steering wheel and Woodruff key. Remove the steering wormshaft front bearing housing and turn the wormshaft forward and out of housing; OR, unbolt worm housing from pedestal and remove wormshaft and housing assembly from tractor. Remove muffler, unbolt hood and lift hood from tractor.

Remove air intake and exhaust pipes and loosen the air cleaner hose clamps. The air cleaner can be removed for convenience. Loosen fan belt and unbolt water pump from lower tank. Remove the cap screws retaining the steering pedestal to tractor and turn the pedestal, as required, to provide clearance for removing the radiator. Unbolt radiator from front end support and using a hoist, lift radiator assembly from tractor.

FAN SHAFT AND VENTILATOR PUMP

All Models

The fan shaft is driven by a bevel pinion which is mounted on the governor shaft. The engine crankcase is ventilated by a rotor-type pump which is mounted on the rear of the fanshaft and bolted to the governor case (Refer to Fig. JD1836).

107. **R&R AND OVERHAUL.** To remove the fan shaft assembly on models without power steering, first remove governor as outlined in paragraph 102 or 103 and proceed as follows: Loosen the air intake hose clamps and remove the air intake casting and the crankcase vent pipe. Remove exhaust pipe. On 620, 630, 720 and 730, remove carburetor, then unbolt and slide manifold forward. On all models, loosen generator and roll generator out toward right side of tractor. Unbolt fan shaft from its front support and on 720 and 730, unbolt the fan shaft front support from tractor frame, being careful not to lose spacer washers which are located between fan support and frame.

On all models, withdraw fan shaft assembly, front end first, from right side of tractor.

NOTE: On models with power steering, the procedure for removing the fan shaft is evident after removing the power steering pump as in paragraph 20.

After fan shaft assembly is removed from tractor, inspect the fan drive bevel gear (70—Fig. JD1837). If the bevel gear must be renewed, refer to paragraph 108.

To disassemble the removed fan shaft, place the assembly in a press and remove the half-moon locks (62) and keeper (61). Remove the assembly from press, disassemble and check the remaining parts against the values which follow:

Fan blade pitch (Inches)
520-530-620-630 .... $2\frac{11}{32}$ - $2\frac{13}{32}$
720-730 .... $2\frac{17}{32}$ - $2\frac{19}{32}$

Thickness of friction washers
All Models .... $\frac{1}{16}$-inch

Friction spring strength
520-530 ...112-138 lbs. @ 1½ inches
620-630-
  720-730 ..216-264 lbs. @ 1½ inches

Bearing take-up spring strength
520-530 ... 70- 86 lbs. @ $1\frac{7}{16}$ inches
620-630-
  720-730 ..171-209 lbs. @ $1\frac{7}{16}$ inches

Ventilator pump specifications are as follows:

Roller diameter 0.4982-0.4988
Roller length 0.965-0.970
Rotor thickness 0.964-0.965
Rotor diameter 2.425-2.427
Rotor groove diameter... 0.500-0.505
Pump body thickness.. 0.9705-0.9745

Use Fig. JD1837 as a general guide during reassembly and be sure to renew all "O" ring packings and felt washers. Tighten the vent pump retaining screws to a torque of 21 Ft.-Lbs.

FAN DRIVE BEVEL GEARS
All Models

The fan drive bevel gears (40 & 70—Fig. JD1838) are available in a matched pair only. Therefore, if either gear is damaged, it will be necessary to renew both gears and adjust them for the proper mesh and backlash.

108. To renew the bevel gears, remove governor as outlined in paragraph 102 or 103 and fanshaft as outlined in paragraph 107. Remove the governor shaft assembly from governor housing and using a suitable puller, remove bevel pinion from governor shaft. Press new bevel pinion on shaft until pinion seats against snap ring. Reassemble governor shaft and reinstall in governor housing.

Using a suitable press, press the fan drive bevel gear (70—Fig. JD1837) further on fan shaft until snap ring (69) can be removed. Remove snap ring and bevel gear and install the new bevel gear.

Before installing fan shaft or governor assembly on tractor, temporarily bolt fanshaft to governor case, observe mesh position of bevel gears and add or remove shim gaskets (71—Fig. JD1838) until heels of gears are in register. Mount a dial indicator in a manner similar to that shown in Fig. JD1839 and check the bevel gear backlash. Add or remove shim gaskets (18—Fig. JD1838) until backlash is 0.004-0.007 for 520, 530, 620 and 630; 0.004-0.006 for 720 and 730.

After proper backlash is obtained, unbolt fan shaft from governor housing and save the mesh adjusting shim gaskets for installation when fan shaft and governor are reinstalled on tractor.

THERMOSTAT
All Models

109. On gasoline and all-fuel models, the thermostat is located in a housing which is bolted to the radiator water inlet casting. On 520, 530, 620 and 630, the removal procedure is evident after an examination of the installation. On

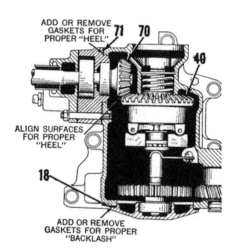

ADD OR REMOVE GASKETS FOR PROPER "HEEL" 71 70

40

ALIGN SURFACES FOR PROPER "HEEL"

18

ADD OR REMOVE GASKETS FOR PROPER "BACKLASH"

Fig. JD1838—Sectional view of a typical governor, showing points for adjusting the fan drive bevel gears.

18. Shims   70. Bevel gear
40. Bevel pinion   71. Shims

Fig. JD1839—Checking backlash of the fan drive bevel gears on 520 & 530 tractors. The same procedure can be used on 620, 630, 720 and 730.

720 and 730 it is possible to renew the unit by merely removing the thermostat housing; however most mechanics prefer to perform the additional work of removing the grille, hood and air cleaner.

110. On LP-Gas models, the thermostat is located in the upper water pipe rear casting. The procedure for removing the thermostat is evident.

WATER PUMP

Coolant leakage at drain hole in pump housing usually indicates a leaking seal.

All Models

111. **R&R AND OVERHAUL.** To remove the water pump on 520, 530, 620 and 630, drain cooling system and disconnect fan belt and lower hose from pump. Disconnect the by-pass line. Remove the two cap screws retaining the fan shaft to the fan shaft front support and unbolt the support from tractor frame. CAUTION: Do not lose shims located between the fan support and tractor frame. Move the fan support rearward, unbolt water pump from radiator and withdraw pump from tractor.

To remove the water pump on 720 and 730, drain cooling system, disconnect fan belt from pump and remove the lower water pipe. Disconnect the bypass line. Unbolt and withdraw water pump from tractor.

On all models, use a suitable puller and remove the drive pulley. Remove retainer ring (90—Fig. JD1840) and press shaft and bearing assembly out of pump housing. Seal (93) can be renewed at this time. The shaft and bearing (91) are available as an assembled unit only.

When reassembling, coat outer surface of seal with a thin coat of shellac and press seal into housing. Install the shaft and bearing unit in pump housing and install retainer ring (90) so that end of ring locks over lug on the housing. Support pump shaft from underneath side and press pulley on shaft until pulley is flush with end of shaft. Place impeller on flat surface and press body and shaft assembly onto impeller until highest vane on impeller is flush with mounting surface of pump body.

NOTE: If any vane protrudes beyond the mounting face of pump housing, the protruding vane will strike the radiator lower tank when pump is installed.

Install pump on tractor by reversing the removal procedure.

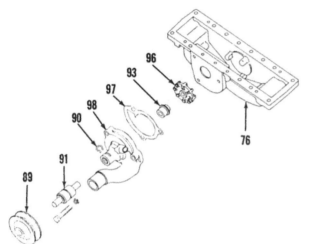

Fig. JD1840 — Exploded view of 520, 530, 620 & 630 water pump. Series 720 & 730 water pump is similarly constructed.

89. Pulley
90. Retainer ring
91. Shaft and bearing assembly
93. Seal
96. Impeller
97. Gasket
98. Pump housing

IGNITION AND ELECTRICAL SYSTEM

DISTRIBUTOR

All Models

112. **APPLICATIONS.** Delco-Remy distributors are used and their applications are as follows:

520-530 1112569
620-630-720-730 1112576

Refer to Appendix I in this manual for distributor test specifications.

113. **INSTALLATION AND TIMING.** The distributor can be timed in either the fully advanced or full retard position. Normally, the distributor is installed and timed in the full retard position; then the running timing is checked and the necessary slight adjustments made at rated engine speed. Breaker contact gap is 0.022.

To install and time the distributor, loosen cap screw and slip cover away from the timing port in flywheel housing. Crank engine until number one (left) piston is coming up on compression stroke and the "TDC" mark on flywheel rim is in register with index in timing port as shown at lower inset in Fig. JD1841.

Install distributor so that spark plug cable terminals in distributor cap are toward governor housing and the rotor arm is under the No. 1 (rear) terminal. Refer to Fig. JD1842. Turn the distributor body clockwise (viewed from above) and turn on the ignition switch. Slowly turn distributor body counter-clockwise until a spark occurs for number one cylinder and tighten the distributor mounting cap screws.

Check the ignition running timing at rated rpm with a timing light as shown in Fig. JD1841. Desired running timing is 20 degrees before top center as shown in upper inset.

Fig. JD1841 — On all models, the static ignition timing is "TDC" and the running timing at rated rpm is 20 degrees before TC.

Fig. JD1842 — Ignition distributor installation on series 620. Other models are similar.

GENERATOR, REGULATOR AND STARTING MOTOR
All Models

114. Delco-Remy electrical units are used and their applications are as follows:

Generator 1100309
Regulator 1118792

Starting Motor:
Series 520 1108155
Series 530 1107725

Series 620 Orchard.
Prior 6223000 1113041

Series 620 Orchard.
After 6222999 1113092

Other Series 620..........1113041

Series 630 1113092

Series 720 prior 7214900....1113304

Series 720 after 7214899....1113079

Series 730 1113093

Refer to Appendix I in this manual for generator, regulator and starting motor test specifications.

On series 620 Orchard tractors after Ser. No. 6222999 and all series 530, 630 and 730 tractors, the starting motor is actuated by Delco-Remy solenoid switches, as follows:

530 1119789
620-630-730 1119778

Refer to Appendix I in this manual for starting motor pinion adjustment.

CLUTCH, BELT PULLEY AND PULLEY BRAKE

CLUTCH ADJUSTMENT
All Models

116. To adjust the clutch, remove the belt pulley dust cover and the cotter pin from each of the three clutch operating bolts. Place the clutch operating lever in the engaged position (lever fully forward) and tighten each

adjusting nut (21—Fig. JD1845) a little at a time and to the same tension. Check tightness of clutch after each adjustment by disengaging and re-engaging clutch. When the adjustment is correct, a distinct snap will occur when the clutch is engaged and 40-80 lbs. will be required at the end of the operating lever to lock the clutch in

the engaged position with engine running at idle speed.

PULLEY BRAKE ADJUSTMENT
Series 520 Prior Ser. No. 5208100
Series 620 Prior Ser. No. 6213100
Series 720 Prior Ser. No. 7214900

116A. Loosen jam nut (10—Fig. JD1846) and back out the adjusting screw (9) until the screw does not contact the pulley brake operating pin

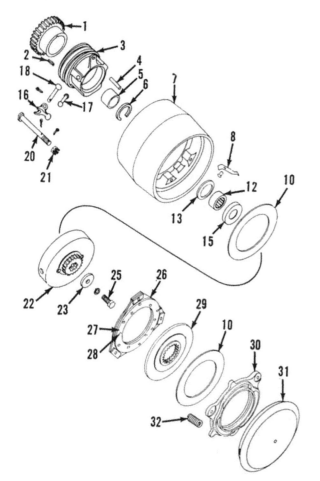

Fig. JD1843—Exploded view of 520 & 530 belt pulley and clutch assembly. The clutch is adjusted with nuts (21).

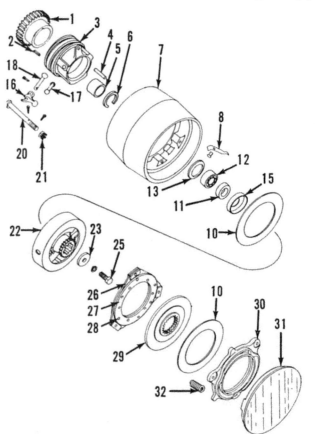

Fig. JD1844—Exploded view of 620 & 630 belt pulley and clutch assembly. The 720 & 730 is similar except two facing discs (26) and two sliding drive discs (29) are used.

1. Drive gear	6. Snap ring	11. Bearing inner	15. Bearing retainer	21. Adjusting nut	28. Rivet

1. Drive gear
2. Key
3. Operating sleeve
4. Drive pin
5. Bushing
6. Snap ring
7. Pulley
8. Spring
10. Clutch facing
11. Bearing inner race (620, 630, 720 & 730)
12. Bearing
13. Bearing washer
15. Bearing retainer
16. Clutch dog
17. Dog toggle
18. Dog pin
20. Operating bolt
21. Adjusting nut
22. Clutch drive disc
23. Washer
26. Facing disc
27. Facing
28. Rivet
29. Sliding drive disc
30. Adjusting disc
31. Pulley dust cover
32. Spring

when clutch lever is pulled back. Then, with the clutch lever in the released position, the belt pulley should rotate freely without any bind.

If binding exists, remove the clutch fork bearing on 520 and 620 or belt pulley assembly and clutch fork on 720. Withdraw the pulley brake operating pin (21—Fig. JD1858 or JD1859) and add one adjusting washer (20) and recheck. When properly adjusted and with the clutch lever in the disengaged position, the pulley should rotate freely and have an end play of $\frac{1}{16}$-$\frac{1}{8}$ inch on the crankshaft. Refer also to Fig. JD1847.

To adjust the pulley brake, pull forward on the brake shoe to insure that the operating pin is contacting the clutch fork. Then while holding the shoe firmly against the pulley, turn the adjusting screw (9—Fig. JD1846) in until it just contacts the operating pin. Turn the screw in ¼-turn more and tighten the jam nut (10).

Series 520 Ser. No. 5208100-5210041—Series 620 Ser. No. 6213100-6217317—Series 720 Ser. No. 7214900-7218699.

116B. Loosen jam nut (10—Fig. JD1846) and back out the adjusting

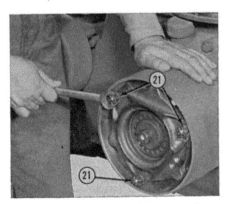

Fig. JD1845—Clutch is adjusted by tightening nuts (21) evenly.

Fig. JD1846—Pulley brake adjusting screw and jam nut on early 520, 620 and 720.

screw (9) until the screw does not contact the pulley brake operating pin when clutch lever is pulled back. Then, with the clutch lever in the released position, the belt pulley should rotate freely without any bind.

If binding exists, remove the pulley brake assembly and shim retainer (14—Fig. JD1858 or JD1859) and remove one adjusting shim (15) and recheck. When properly adjusted and with the clutch lever in the disengaged position, the pulley should rotate freely and have an end play of $\frac{1}{16}$-$\frac{1}{8}$ inch on the crankshaft. Refer also to Fig. JD1848.

To adjust the pulley brake, pull forward on the brake shoe to insure that the operating pin is contacting the clutch fork. Then while holding the shoe firmly against the pulley, turn the adjusting screw (9—Fig. JD1846) in until it just contacts the operating pin. Turn the screw in ¼-turn more and tighten the jam nut (10).

Series 520 After Ser. No. 5210041
Series 620 After Ser. No. 6217317
Series 720 After Ser. No. 7218699
Series 530-630-730

116C. These late production tractors can be identified in that they have two adjusting screws in the pulley brake as shown in Fig. JD1849. This late construction differs from both early constructions as follows:

The fork stop has been removed from the pulley brake operating pin and adjustable stop screws are located between clutch collar and housing. Pivot pin between the clutch collar and the clutch fork is located in a slot and spring loaded. The clutch lever now moves into a neutral position when the clutch is disengaged and further rearward movement of the clutch lever compresses the springs and moves the pivot pin in the slot to engage the pulley brake. A limit screw

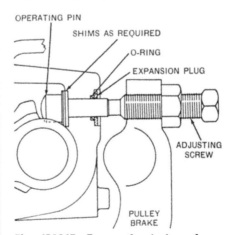

Fig. JD1847—Top sectional view of very early production 520, 620 & 720 clutch fork bearing, showing the installation of shim adjusted pulley brake operating pin.

has been added to the brake and serves as a stop to prevent the pulley brake spring from holding the clutch in partial engagement when the lever is in neutral.

On 520 and 530 tractors, the adjustable stop screws are located in the right main bearing housing, and the stops are machined on the clutch collar. Access to the screws is obtained by removing the clutch fork shaft bearing; then separate the clutch fork and collar and insert the collar in the operating sleeve. Proceed to paragraph 116D.

On 620 and 630, the adjustable stop screws are located in the main bearing housing and the stops are located on the clutch collar. Access to the screws is obtained through pipe plug openings in top and bottom of the reduction gear cover. Proceed to paragraph 116D.

On 720 and 730, the adjustable stop screws are located in the clutch collar and the stop pads are machined on the

Fig. JD1848—On some 520, 620 and 720 models, the pulley brake operating pin adjusting shims are externally located as shown.

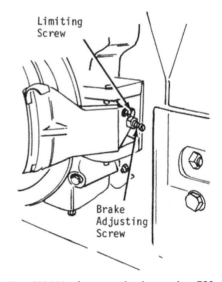

Fig. JD1849—Late production series 720 pulley brake and adjusting screws. The adjusting screws are similarly located on late 520 and 620 and all 530, 630 and 730.

main bearing housing. Access to the screws is obtained through rectangular openings in top and bottom of reduction gear cover. Proceed to the following paragraph.

116D. Before making the pulley brake adjustment, make certain that the crankshaft end play is properly adjusted as in paragraph 60 and the clutch is properly adjusted as in paragraph 116.

With the clutch disengaged, use a large screw driver or pry bar as shown

Fig. JD1850—Using screw driver to hold the belt pulley outward.

in Fig. JD1850 and hold the pulley out against the fixed clutch drive disc. Working through the previously mentioned access openings, adjust the clearance between the stop screws and the machined stop bosses or pads to 0.060. The procedure for checking the adjustment on 720 and 730 is shown in Fig. JD1851 where the pulley is removed for illustrative purposes only. On 720 and 730 reinstall the access opening covers. On 620 and 630, use Permatex on the pipe plugs and reinstall the plugs. On 520 and 530, reassemble the clutch fork and collar and reinstall the clutch fork bearing.

With the clutch lever in neutral position, turn the brake adjusting screw (Fig. JD1849) inward until the pulley brake just touches the pulley and a slight drag is felt when pulley is rotated. Tighten the jam nut. Engage the clutch and adjust the limiting screw until there is $\frac{1}{32}$-inch clearance between the pulley and the pulley brake facing. Tighten the jam nut.

RENEW CLUTCH FACINGS
All Models

117. To remove the clutch discs and facings, remove the pulley dust cover and adjusting disc (30—Figs. JD1843 or 1844). Withdraw the lined and un-

lined discs. Remove cap screw (25) retaining the clutch drive disc (22) to crankshaft and using jack screws or a suitable puller as shown in Figs. JD1856 and JD1857, remove the clutch drive disc and inner facing.

Worn, badly glazed or oil soaked facings should be renewed. A facing that is in usable condition, is quite rigid. Any facing that bends easily should be renewed. Renew release springs (32—Fig. JD1843 or 1844) if they are rusted, distorted or do not meet the following specifications:

Pounds Test @ Height

520-530	20-30 lbs. @ $1\frac{5}{16}$ inches
620-630	45-55 lbs. @ $1\frac{5}{8}$ inches
720-730	45-55 lbs. @ $1\frac{5}{8}$ inches

Use Fig. JD1843 or JD1844 as a guide during reassembly and install the clutch drive disc so that "V" mark on drive disc is in register with flat spot at end of one of the crankshaft splines. Adjust the clutch as outlined in paragraph 116.

R&R BELT PULLEY
Series 520-530-620-630

118. To remove the belt pulley, first remove the clutch facings as outlined in paragraph 117 and disconnect the clutch operating rod from the clutch fork shaft. Unbolt and remove the pul-

Fig. JD1851 — Checking the clutch fork stop screw adjustment on 720 & 730. The belt pulley is removed for illustrative purposes only. The procedure for checking other models is similar.

Fig. JD1856 — Using a suitable puller to remove the clutch drive disc on 520 & 530.

Fig. JD1857 — Using ½-inch jack screws to remove the clutch drive disc on 620, 630, 720 and 730.

ley brake, clutch fork shaft, bearing and clutch fork as an assembly from tractor. Withdraw pulley from crankshaft.

Series 720-730

119. To remove the belt pulley, first remove the clutch facings as outlined in paragraph 117 and disconnect the clutch operating rod from the clutch fork shaft. Remove pivot pin and withdraw pulley brake (11—Fig. JD1859). Unbolt cover (26) from reduction gear case and withdraw pulley assembly from tractor.

OVERHAUL PULLEY

All Models

120. To disassemble the removed pulley, use a punch and hammer as shown in Fig. JD1860 and remove the bearing retainer and bearing (12—Fig. 1843 or 1844). Remove the operating bolts (20), pins (18), dogs (16) and toggles (17). Using a suitable puller, remove drive gear (1) and slide operating sleeve from pulley.

Check pulley bushing (5) and the engine crankshaft against the values which follow:

Crankshaft diameter at pulley bushing
520-530 ....2.1815-2.1825
620-630-720-730 ......2.432 -2.434

Pulley bushing inside diameter (Minimum)
520-530 ....2.186
620-630-720-730 ....2.4375

Suggested clearance between pulley bushing and crankshaft
520-530 ....0.0035-0.0045
620-630-720-730 ....0.0035-0.0055

If clearance between pulley bushing (5) and crankshaft is excessive, renew the bushing. Using a piloted drift, press bushing into pulley until bushing seats against snap ring (6). If bushing is carefully installed, no final sizing will be required.

When reassembling, heat gear (1) to approximately 300 deg. F., install gear with long hub toward pulley and lubricate toggles (17) as follows:

Any one of the following special lubricants are recommended by John Deere for lubricating the clutch toggles:

Calumet Viscous Lubricant, 10X, manufactured by Standard Oil Company of Indiana.
No. 102 Cosmolube, manufactured by E. F. Houghton and Co., 303 West Lehigh Ave., Philadelphia 33, Penn.

On 520 and 530 and early production

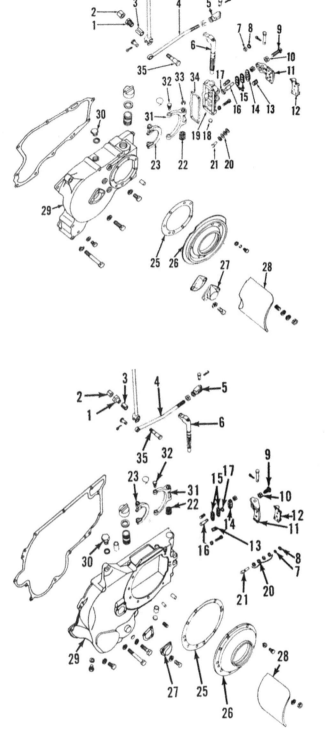

Fig. JD1858 — Series 520 prior Ser. No. 5210042 and 620 prior Ser. No. 6217318 clutch and belt pulley operating parts exploded from the first reduction gear cover. *Parts used on series 520 prior 5208100 and series 620 prior 6213100. **Parts used on series 520 Ser. No. 5208100-5210041 and series 620 Ser. No. 6213100-6217317.

1. Nut, special
2. Nut
3. Bushing
4. Operating rod
5. Yoke
6. Clutch fork shaft
7. "O" ring*
8. Expansion plug*
9. Adjusting screw
10. Jam nut
11. Pulley brake
12. Brake lining
13. Spring
14. Operating pin stop**
15. Steel shims**
16. Operating pin**
17. "O" ring**
18. Expansion plug
19. Clutch fork bearing
20. Shim washer*
21. Operating pin*
22. Fork spring
23. Clutch collar
25. Gasket
26. Dust shield
27. Cover
28. Pulley guard
29. First reduction gear cover
30. Plug
31. Clutch fork
32. Fork pivot
33. Snap ring
34. Gasket
35. Pivot bolt

Fig. JD1859—Series 720 (prior to 7218700) clutch and belt pulley operating parts exploded from the first reduction gear cover. *Parts used on models prior to 7214900. **Parts used 7214899 through 7218699.

series 620 and 720, pack the recess in end of each clutch toggle with lubricant and install lubricated end of toggle into sockets in the clutch operating sleeve. Lubricate opposite end of each toggle with the same special lubricant.

On 630 and 730 as well as late production 620 and 720, lubricate end of toggle which goes into the clutch dog cup with one of the special lubricants. After pulley is assembled, place the operating sleeve in the engaged position (away from pulley gear) and force the special grease into the grease fitting until grease appears around toggle ends. This procedure fills the grease reservoir in the operating sleeve and provides toggle lubrication via small holes to the toggle sockets.

Note: Early 620 and 720 tractors may be fitted with current production clutch operating sleeve which incorporates a grease fitting.

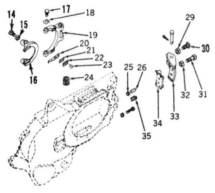

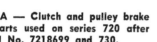

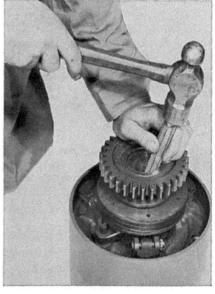

Fig. JD1859B — Clutch and pulley brake operating parts used on series 520 after Serial No. 521-0041 and 530. Series 620 after Serial No. 6217317 and 630 are similar.

Fig. JD1859A — Clutch and pulley brake operating parts used on series 720 after Serial No. 7218699 and 730.

14. Limit screw	21. Reinforcing spring
15. Jam nut	(720 & 730)
16. Clutch collar	22. Plate (720 & 730)
17. Fork pivot	23. Rivet (720 & 730)
18. Lock plate (720	24. Fork shaft spring
& 730)	25. "O" ring
19. Fork	26. Pulley brake operating
20. Spring	pin

27. Expansion plug	33. Pulley brake
(520, 530, 620, 630)	34. Brake lining
28. Fork bearing	35. Spring
(520, 530, 620, 630)	36. Gasket (520, 530,
29. Jam nut	620 & 630)
30. Limit screw	37. Snap ring (520, 530,
31. Brake adjusting screw	620 & 630)
32. Jam nut	

38. Roll pin (520, 530,	
620 & 630)	
39. Pin (520, 530,	
620, 630)	
40. Expansion plug	
(520, 530, 620, 630)	

CLUTCH CONTROLS

Series 520-530-620-630

121. **R&R AND OVERHAUL.** On early models of the 520 and 620 series, the clutch fork (31—Fig. JD1858) and clutch collar (23) can be renewed after disconnecting the clutch operating rod, unbolting the clutch fork shaft bearing from the reduction gear cover and withdrawing bearing and fork assembly from tractor. The procedure for subsequent disassembly is evident. Be sure to check all parts thoroughly and renew any which are damaged or show signs of wear. Series 530 and 630 as well as late series 520 and 620 are serviced in a similar manner and the differing clutch operating parts are shown in Fig. JD1859B.

Series 720-730

122. **R&R AND OVERHAUL.** On early models of the 720 series, the clutch fork (31—Fig. JD1859) and clutch collar (23) can be renewed after removing the belt pulley as outlined in paragraph 119. The procedure for subsequent disassembly is evident. Be sure to check all parts thoroughly and renew any which are damaged or show signs of wear. Series 730 as well as late series 720 are serviced in a similar manner and the differing clutch operating parts are shown in Fig. JD1859A.

Fig. JD1860—Using a punch and hammer to remove the pulley bearing.

TRANSMISSION

OVERHAUL
Series 520-530

123. **FIRST REDUCTION GEAR COVER.** To remove the first reduction gear cover, disconnect the clutch operating rod and drain oil from cover. Remove belt pulley as outlined in paragraph 118. Move the right rear wheel out and remove the right brake assembly as outlined in paragraph 152. Remove the transmission drive shaft right bearing cover (27—Fig. JD1858) from reduction gear cover, extract cotter pin and remove nut (49—Fig. JD1873) from end of shaft. Remove cap screws retaining the reduction gear cover to main case and bump

end of transmission drive shaft with a soft hammer to loosen the reduction gear cover. When cover is free from dowels, withdraw cover, front end first, from main case.

When reassembling, soak new reduction gear cover gasket until gasket is pliable, shellac gasket to main case and install reduction gear cover by reversing the removal procedure. Pour about 1 quart of transmission oil into the reduction gear cover.

124. **TRANSMISSION TOP COVER.** To remove the transmission top cover and shifter quadrant assembly, proceed as follows: Disconnect battery cable and oil gage line. Remove the

four cap screws retaining instrument panel to the steering shaft rear support, pull instrument panel rearward and disconnect wires which go through the gear shifter quadrant. Remove both cap screws retaining the steering shaft rear support to the gear shifter quadrant, raise steering shaft support and rear end of hood approximately 1½-inches and block up between hood and governor housing. Remove the cap screws retaining the shifter quadrant and transmission cover assembly to main case and with gear shift lever on right side of quadrant, withdraw top cover and shifter quadrant assembly from tractor.

Fig. JD1861 — Series 720 prior to Serial No. 721-8700 clutch fork (31) and collar (23) installation as viewed after pulley has been removed.

125. SHIFTER SHAFTS AND SHIFTERS. To remove the shifter shafts and shifters, first remove the engine flywheel as outlined in paragraph 59 and the transmission top cover as outlined in paragraph 124. The fourth and sixth speed shifter shaft pawl and spring should be removed at this time. They are located in a vertical drilled hole in main case to the left of the top opening and are retained by the transmission top cover. Using a pry bar through transmission top opening, move each shifter along its shaft until the detent pawls rise and hold the pawls in the raised position by inserting a cotter pin or wire in the exposed hole in each pawl. Un-

wire and remove set screw (26—Fig. JD1871) which positions the fourth and sixth speed shifter on its shaft.

Remove the fourth and sixth speed gear cover which is located under flywheel on left side of main case. Remove cap screw (18—Fig. JD1871 or 1872) and adjusting screw (21) from left end of each shifter shaft. Pull each shifter shaft toward left to disengage it from locking plate (30 — Fig. JD1871) which holds the right end in place, rotate shafts sufficiently to move detents out of alignment with pawls, withdraw shifter shafts from left and shifters from above. Remove the rear shifter shaft and shifter first, then work forward, removing the remaining shafts and shifters.

New shifter yokes can be riveted to shifters if old yokes are worn or bent. Renew any shifter shaft that is worn around the detent. Renew any pawl that is worn out-of-round at ball end. The first and third as well as the second and fifth speed shifter springs (34) should test 42-51 lbs. when compressed to 1¾-inches. The fourth and sixth speed shifter spring should have a free length of 1-1¼-inches.

When reinstalling the shifter shafts and shifters, refer to Fig. JD1871 and

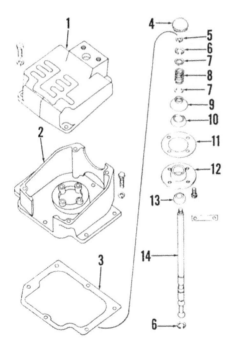

Fig. JD1870—Exploded view of 520 & 530 shifter quadrant and transmission cover assembly.

1. Quadrant	8. Spring
2. Cover	9. Ball socket cover
3. Gasket	10. Fulcrum seal
4. Shift ball	11. Gasket
5. Lock washer	12. Ball socket
6. Snap ring	13. Fulcrum ball
7. Washer	14. Gear shift lever

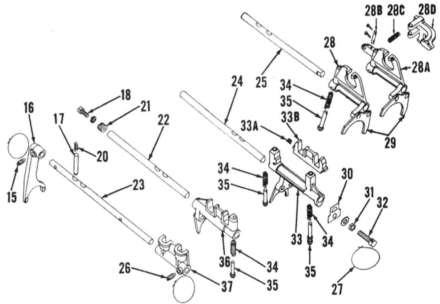

Fig. JD1871—Exploded view of 520 & 530 shifter shafts and shifters. Stop screw (32) and lock plate (30) are accessible after removing the first reduction gear cover.

15. Set screw	23. Fourth and sixth shifter shaft	28B. Pin, after 5203321	33A. Spring
16. Fourth & sixth speed shifter fork	24. First, third and reverse shifter shaft	28C. Spring, after 5203321	33B. First, third and reverse shifter gate
17. Fourth & sixth pawl	25. Underdrive shifter shaft	28D. Underdrive shifter lock, after 5203321	34. Pawl spring (three used)
18. Locking cap screw (three used)	26. Drilled set screw	29. Underdrive shifter yoke	35. Pawl (three used)
20. Pawl spring	28. Underdrive shifter, prior 5203322	30. Lock plate	36. Second and fifth speed shifter
21. Adjusting screw (three used)	28A. Underdrive shifter, after 5203321	32. Stop screw	37. Fourth and sixth speed shifter
22. Second and fifth shifter shaft		33. First, third and reverse shifter	

reverse the removal procedure, making certain that flat on right end of shifter shafts engage locking plate (30). After the shifter shafts and shifters are installed, place the fourth and sixth speed shifter in neutral position, making certain that pawl engages detent in shaft. Turn the adjusting screw (21), located at left end of shafts (22, 24 and 25), in or out, until left hand shifter gates are aligned. After the adjustment is complete, install and tighten the locking cap screws (18) securely.

Move shifter (33) to first or third speed position; at which time, there should be a gap of approximately 5/64 inch between end of stop screw (32) and right end of shifter (33), and the first and third speed drive and sliding gears should be meshed properly.

NOTE: If stop screw (32) is screwed in far enough to prevent first and third speed gears from going into full mesh or if gap between stop screw and shifter is excessive enough to permit more than ⅛-inch overshift, it will be necessary to remove clutch, belt pulley and reduction gear cover as in paragraph 123 to permit readjustment of the stop screw.

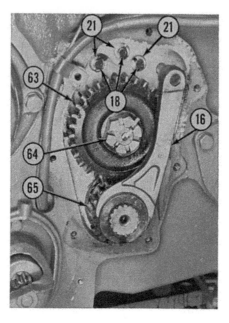

Fig. JD1872—Left side view of 520 & 530 main case with fourth and sixth speed gear cover removed.

16. Fourth and sixth speed shifter fork	63. Fourth and sixth drive gear
18. Locking cap screws	64. Nut
21. Adjusting screws	65. Fourth and sixth speed pinion

126. SLIDING GEAR SHAFT. To remove the sliding gear shaft (57—Fig. JD1873) and gears, remove the transmission top cover as in paragraph 124 and the shifter shafts and shifters as outlined in paragraph 125. Remove cotter pin and nut (64—Fig. JD1872) from left end of sliding gear shaft and using a suitable puller, remove the fourth and sixth speed drive gear (63). Remove the sheet metal oil retainer (62—Fig. JD1873), extract snap ring (59) and pull the sliding gear shaft toward left until the left bearing (60) emerges from the main case. Withdraw gearshaft from left side of case and remove gears from above. Pilot bearing (53) can be removed from the drive shaft after extracting snap ring (53B).

Inspect all parts and renew any which are questionable. Install bearing (53), washer (53A) and snap ring (53B). Install sliding gear shaft and gears by reversing the removal procedure and install snap ring (59) with gap in snap ring spanning the oil passage in main case. Install the sheet metal oil retainer (62) with flat spot adjacent to oil passage in main case.

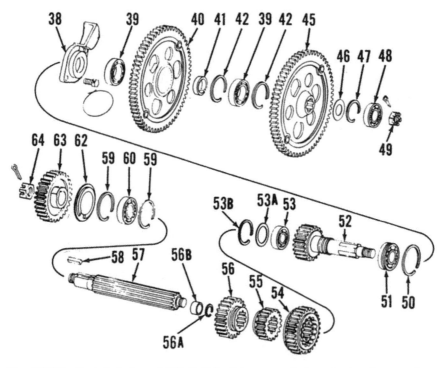

Fig. JD1873—Exploded view of 520 & 530 sliding gear shaft and related parts. Gear (40) is the engine driven powershaft idler. Late models are fitted with a Woodruff key instead of key (58).

38. Bearing cover	51. Bearing	56. Second and fifth sliding pinion
39. Bearing	52. Transmission drive shaft	56A. Snap ring
40. Powershaft idler gear	53. Pilot bearing outer race with rollers	56B. Bearing inner race
41. Spacer		57. Sliding gear shaft
42. Snap ring	53A. Washer	58. Key
45. First reduction gear	53B. Snap ring	59. Snap ring
46. Washer	54. Sliding gear shaft drive gear	60. Bearing
47. Snap ring		62. Oil retainer
48. Bearing	55. First and third sliding pinion	63. Fourth and sixth drive gear
49. Nut		
50. Snap ring		

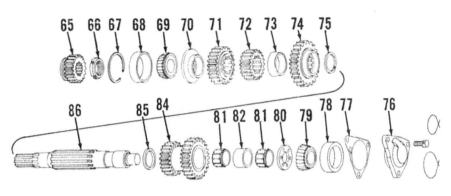

Fig. JD1874—Exploded view of series 520 & 530 transmission countershaft and gears.

65. Fourth and sixth speed sliding pinion	71. Second and fifth speed gear	77. Shims & gaskets
66. Lock nut	72. Differential drive pinion	78. Bearing cup
67. Snap ring	73. Spacer	79. Bearing cone
68. Bearing cup	74. First and third speed gear	80. Thrust washer
69. Bearing cone	75. Snap ring	81. Bearing
70. Spacer	76. Bearing housing	82. Spacer
		84. Idler gear
		85. Collar
		86. Countershaft

After installing the fourth and sixth speed drive gear, tighten nut (64) securely.

127. TRANSMISSION DRIVE SHAFT. To remove the drive shaft (52—Fig. JD1873), remove the clutch, belt pulley and first reduction gear cover as outlined in paragraph 123 and the sliding gear shaft as outlined in the preceding paragraph 126. Withdraw the first reduction gear (45) and the power shaft idler gear (40). Remove bearing cover (38) and withdraw drive shaft from tractor. Pilot bearing (53) can be removed from the drive shaft after extracting snap ring (53B) and removing washer (53A).

Fig. JD1875—Using dial indicator to check end play of the 520 & 530 transmission countershaft. The recommended end play of 0.001-0.004 is controlled by shims under the right bearing housing.

When installing the shaft, reverse the removal procedure.

128. COUNTERSHAFT. To remove the countershaft (86—Fig. JD1874) remove the sliding gear shaft as in paragraph 126 and the transmission drive shaft as outlined in the preceding paragraph. Remove the countershaft right bearing housing (76) and unstake and remove nut (66) from left end of shaft. Using a soft hammer, bump countershaft out right side of case and remove gears from above.

The countershaft right bearing cup can be pulled from bearing housing (76) if renewal is required and the left bearing cup can be driven from the main case bore.

When reassembling, use Fig. JD1874 as a guide, install the same number of shims (77) as were originally removed and tighten nut (66) securely. Stake the nut into one of the shaft splines. Mount a dial indicator as shown in Fig. JD1875 and check the shaft end play which should be 0.001-0.004. If end play is not as specified, remove the countershaft right bearing housing (76—Fig. JD1874) and add or remove the required amount of shims (77).

Series 620-630

The following procedures do not specifically cover the 620 orchard tractors; however, orchard models can be serviced in a similar manner by using the included exploded views of the orchard transmission components.

130. FIRST REDUCTION GEAR COVER. To remove the first reduction gear cover, disconnect the clutch operating rod and drain oil from cover. Remove belt pulley as outlined in paragraph 118. Move the right rear wheel out and using a double nut arrangement, remove the upper right hand implement mounting stud from the right front face of the rear axle housing. Unbolt and withdraw the

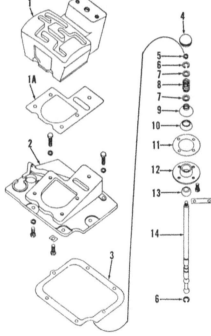

Fig. JD1876—Exploded view of series 620 (except orchard) & 630 transmission top cover and shifter quadrant assembly.

1. Quadrant	8. Spring
2. Transmission cover	9. Ball socket cover
3. Gasket	10. Ball socket seal
4. Gear shift ball	11. Gasket
5. Lock washer	12. Fulcrum ball socket
6. Snap ring	13. Fulcrum ball
7. Washer	14. Gear shift lever

Fig. JD1876A — Exploded view of series 620 orchard transmission case cover and shifter quadrant assembly.

1. Shifter ball
2. Gear shift lever
3. Woodruff key
4. Rod
5. Quadrant
6. Top cover
7. Arm
8. Gasket

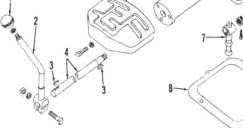

right brake assembly. Remove the transmission drive shaft right bearing cover (27—Fig. JD1858), extract cotter pin and remove nut (44—Fig. JD1880) from end of shaft. Remove cap screws retaining the reduction gear cover to main case and bump end of transmission drive shaft with a soft hammer to loosen the reduction gear cover. Pry cover from its locating dowels and withdraw the first reduction gear cover from tractor. Be careful not to lose the camshaft end play removing spring which is retained by the reduction gear cover.

When reassembling, soak new reduction gear cover gasket until gasket is pliable, shellac gasket to main case and install reduction gear cover by reversing the removal procedure. Pour about 1½-quarts of transmission oil into the reduction gear cover.

131. TRANSMISSION TOP COVER. To remove the transmission top cover and shifter quadrant assembly, remove grille and loosen the two cap screws retaining front of hood to the radiator top tank. Remove the four cap screws retaining instrument panel to support, disconnect oil pressure line from gage, pull instrument panel rearward and remove the two cap screws retaining the steering shaft and hood support to the gear shift quadrant. Disconnect battery cable and disconnect wires from instrument panel which go through the gear shifter quadrant. Disconnect oil lines

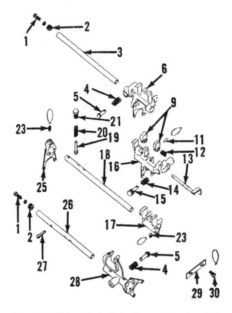

Fig. JD1878—Exploded view of series 620 orchard transmission shifter shafts and shifters.

1. Locking cap screw	18. Fourth and sixth shifter shaft
2. Adjusting screw	19. Fourth and sixth shifter shaft
3. Shifter shaft	20. Fourth and sixth pawl spring
4. Shifter pawl spring	21. Retainer
5. Shifter pawl	23. Set screw
6. First, third and reverse shifter	25. Fourth and sixth speed shifter arm
9. Shifter arm	26. Overdrive shifter shaft
11. Roll pin	27. Cotter pin
12. Spring	28. Overdrive shifter
13. Shifter lock	29. Shifter shaft lock plate
14. Pawl spring	30. Cap screw
15. Shifter pawl	
16. Second and fifth shifter	
17. Fourth and sixth shifter	

from the automatic fuel shut off located at top of fuel filter and disconnect oil line from connector located at right rear of governor case. Disconnect speed control rod from governor spring, fuel line from carburetor, coil wire from coil and fuel tank support from governor housing. Raise rear of hood approximately 2 inches and block up between hood and governor housing. Remove the cap screws retaining the shifter quadrant and transmission cover assembly to main case and withdraw assembly from tractor.

132. SHIFTER SHAFTS AND SHIFTERS. To remove the shifter shafts and shifters, first remove the engine flywheel as outlined in paragraph 59 and the transmission top cover as outlined in the preceding paragraph 131. Remove the fourth and sixth speed shifter pawl and spring. They are located in a vertical drilled hole in main case to the left of the top opening and are retained by threaded retainer (32—Fig. JD1877).

Using a pry bar through transmission top opening, move each shifter along its shaft until the detent pawls rise and hold the pawls in the raised position with a cotter pin inserted in the exposed hole of each pawl.

Move the left rear wheel out on the axle and remove the fourth and sixth speed gear cover which is located under flywheel on left side of main case.

Fig. JD1877 — Exploded view of series 620 (except orchard) and 630 shifter shafts and shifters.

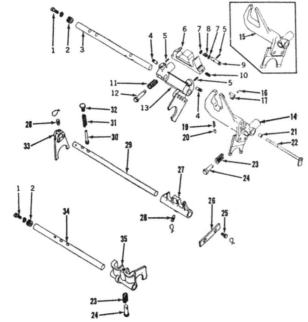

1. Locking cap screw	
2. Adjusting screw	
3. Second and fifth shifter shaft	
4. Pin	
5. Roll pin	
6. Second and fifth shift gate	
7. Washer	
8. Spring	
9. Lock pin	
10. Spring	
11. Shifter pawl spring	
12. Shifter pawl	
13. Second and fifth shifter	
14. Overdrive shifter, after 6211593	
15. Overdrive shifter, prior 6211594	
16. Roll pin, after 6211593	
17. Overdrive shifter rod arm, after 6211593	
19. Overdrive shifter rod pin spring, after 6211593	
20. Overdrive shifter rod pin, after 6211593	
21. Bushing, after 6211593	
22. Overdrive shifter rod, after 6211593	
23. Shifter pawl spring	
24. Shifter pawl	
25. Cap screws	
26. Lock plate	
27. Fourth and sixth shifter	
28. Set screw	
29. Fourth and sixth shifter shaft	

30. Fourth and sixth shifter pawl	33. Fourth and sixth shifter arm
31. Fourth and sixth shifter pawl spring	34. First, third and reverse shifter shaft
32. Retainer	35. First, third and reverse shifter

Fig. JD1879—Left side of 620 & 630 main case with fourth and sixth speed gear cover removed.

1. Locking cap screw	45. Fourth and sixth speed sliding gear
2. Adjusting screw	
28. Set screw	62. Fourth and sixth speed drive gear
33. Fourth and sixth speed shifter	

Remove cap screw (1—Figs. JD1877 and 1879) and adjusting screw (2) from left end of front and rear shifter shaft. Pull each shifter shaft toward left to disengage it from locking plate (26—Fig. JD1877), rotate shafts sufficiently to move detents out of alignment with pawls, withdraw shifter shafts from left and shifters from above.

New shifter yokes can be riveted to shifters if old yokes are worn or bent. Renew any shifter shaft that is worn around the detents. Renew any pawl that is worn out-of-round at ball end. The fourth and sixth speed shifter spring should test 51-63 pounds when compressed to 1⅞-inches.

When reinstalling the shifter shafts and shifters, refer to Fig. JD1877 and reverse the removal procedure, making certain that flat on right end of shifter shafts engage locking plate (26). After the shifter shafts and shifters are installed, place the fourth and sixth speed shifter in neutral position,

making certain that pawl (30) engages detent in shaft. Turn adjusting screw (2), located at left end of front and rear shifter shaft, until right hand gates are aligned with gate of shifter on center shaft.

133. SLIDING GEAR SHAFT. To remove the sliding gear shaft (56—Fig. JD1880) and gears, remove the transmission case top cover as in paragraph 131 and the shifter shafts and shifters as outlined in paragraph 132. Remove the sheet metal oil retainer (46) and extract outer snap ring (48) from main case. Withdraw sliding gear shaft from left side of main case and remove gears from above. Pilot bearing (52) can be removed from the transmission drive shaft after removing snap ring (51).

Use Fig. JD1880 as a guide during installation and install the sliding gear shaft by reversing the removal procedure. Install outer snap ring (48) with gap in snap ring spanning the oil

passage in main case. Install the sheet metal oil retainer (46) with flat spot adjacent to oil passage in main case.

134. TRANSMISSION DRIVE SHAFT. To remove the drive shaft (53—Fig. JD1880), remove the clutch, belt pulley and first reduction gear cover as outlined in paragraph 130. Remove the sliding gear shaft as in the preceding paragraph 133.

Withdraw the first reduction gear (40) and powershaft idler gear (37). Remove bearing cover (35) and withdraw drive shaft from tractor.

When installing the shaft, reverse the removal procedure.

135. COUNTERSHAFT. To remove the transmission countershaft (72—Fig. JD1881), remove the sliding gear shaft as in paragraph 133 and the transmission drive shaft as in the preceding paragraph. Remove cotter pin and nut (60) from left end of countershaft and using a suitable puller, remove the fourth and sixth speed drive gear (62). Remove the countershaft

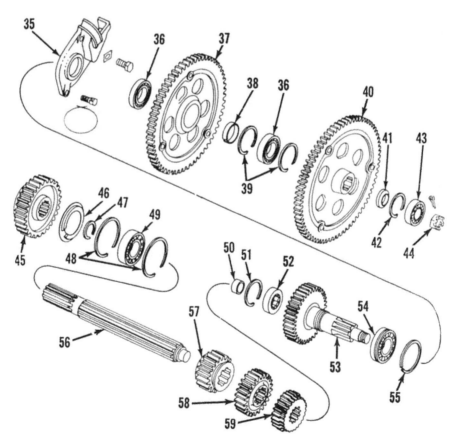

Fig. JD1880—Exploded view of 620 & 630 sliding gear shaft, transmission drive gear and associated parts.

35. Bearing cover	45. Fourth and sixth speed sliding gear	53. Transmission drive shaft
36. Bearing	46. Oil retainer	54. Bearing
37. Powershaft idler	47. Snap ring	56. Sliding gear shaft
38. Spacer	48. Snap ring	57. First and third sliding pinion
39. Snap ring	49. Bearing	58. Second and fifth sliding pinion
40. First reduction gear	50. Pilot bearing inner race	59. Sliding gear shaft drive gear
41. Spacer	51. Snap ring	
42. Snap ring	52. Pilot bearing	
43. Bearing		
44. Nut		

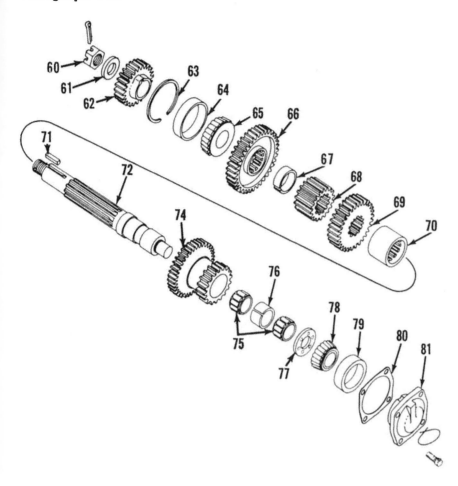

Fig. JD1881—Exploded view of 620 & 630 transmission countershaft. Recommended end play of 0.001-0.004 is adjusted with shims and gaskets (80).

60. Nut	67. Spacer	74. Countershaft idler
61. Washer	68. Differential drive	gear
62. Fourth and sixth	pinion	75. Bearing
speed drive gear	69. Second and fifth speed	76. Spacer
63. Snap ring	gear	77. Thrust washer
64. Bearing cup	70. Spacer	78. Bearing cone
65. Bearing cone	71. Key	79. Bearing cup
66. First and third speed	72. Countershaft	80. Shims and gaskets
gear		81. Bearing housing

Fig. JD1882 — Using a dial indicator to check end play of 620 & 630 transmission countershaft. Recommended end play of 0.001-0.004 is controlled by shims under the right bearing housing.

60. Nut	speed drive gear
62. Fourth and sixth	72. Countershaft

When reassembling, soak new reduction gear cover gasket until gasket is pliable, shellac gasket to main case and install reduction gear cover by reversing the removal procedure. Make certain that oiler gear (47—Fig. JD1883) meshes properly with the first reduction gear before tightening the cover cap screws. Pour about 1½-quarts of transmission oil into reduction gear cover.

right bearing housing (81), bump countershaft out right side of main case and withdraw gears from above.

The right bearing cup can be pulled from bearing housing (81) if renewal is required and the left bearing cup can be driven from the main case bore.

Use Fig. JD1881 as a guide during reassembly and install the same number of shims (80) as were originally removed. Install the fourth and sixth speed drive gear (62) and tighten nut (60) securely. Mount a dial indicator as shown in Fig. JD1882 and check the countershaft end play which should be 0.001-0.004. If the end play is not as specified, remove the countershaft right bearing housing (81 — Fig. JD1881) and add or remove the required amount of shims (80).

Series 720-730

136. **FIRST REDUCTION GEAR COVER.** To remove the first reduction gear cover, disconnect the clutch operating rod and drain oil from cover. Remove belt pulley as outlined in paragraph 119. Move the right wheel out, unbolt and remove the right brake assembly. Remove the transmission drive shaft right bearing cover, extract cotter pin and remove nut from end of transmission drive shaft. Remove cap screws retaining the reduction gear cover to main case and bump end of drive shaft with a soft hammer to loosen the reduction gear cover. Pry cover from its locating dowels and withdraw the first reduction gear cover from tractor. Be careful not to lose the camshaft end play removing spring which is retained by the reduction gear cover.

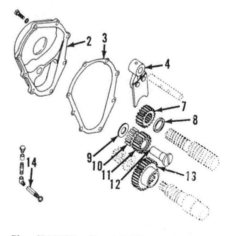

Fig. JD1882A—Special 2½ m.p.h. reverse speed gear attachment for series 620 and 630 tractors.

2. Reverse gear	9. Thrust washer
cover	10. Rollers
3. Gasket	11. Idler gear
4. Reverse gear	12. Idler gear shaft
shifter arm	13. Reverse gear
7. Sliding pinion	14. Oil level exten-
8. Spacer	sion connector

137. TRANSMISSION TOP COVER. To remove the transmission top cover and shifter quadrant assembly, remove grille and loosen the two cap screws retaining front of hood to the radiator top tank. Remove the four cap screws retaining instrument panel to support, disconnect oil pressure line from gage, pull instrument panel rearward and remove the two cap screws retaining the steering shaft and hood support to the gear shift quadrant. Disconnect battery cable and disconnect wires which go through the gear shifter

quadrant from instrument panel. Disconnect oil lines from the automatic fuel shut off valve located at top of fuel filter and disconnect oil line from connector located at right rear of governor case. Disconnect speed control rod from governor spring, fuel line from carburetor, coil wire from coil and fuel tank support from governor housing and upper water pipe. Raise rear of hood approximately 2 inches and block up between hood and governor housing. Remove tool box. Remove the cap screws retaining the shifter quadrant and transmission cover assembly to main case and withdraw assembly from tractor.

When installing the cover, make certain that end of gearshift lever enters the shifter gates at left side of shifters.

138. SHIFTER SHAFTS AND SHIFTERS. To remove the transmission shifter shafts and shifters, first remove the engine flywheel as outlined in paragraph 59 and the transmission top cover as in the preceding paragraph. Move the left wheel out and remove the gear cover which is located on left side of main case. On models prior to Ser. No. 7202781 slide overdrive shifter (8A—Fig. JD1885) along shaft (3) until the detent pawl rises and hold the pawl in

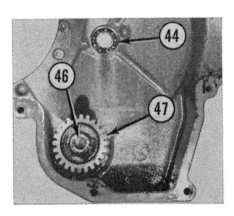

Fig. JD1883—Inside view of 720 & 730 reduction gear cover, showing the installation of the oil slinger gear.

44. Transmission drive shaft right bearing
46. Oil slinger gear pin
47. Oil slinger gear

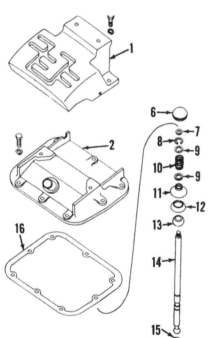

Fig. JD1884—720 & 730 transmission cover and shifter quadrant assembly.

1. Quadrant
2. Transmission cover
7. Lock washer
8. Snap ring
9. Washer
10. Spring
11. Fulcrum ball socket cover
12. Ball socket seal
13. Fulcrum ball
14. Shift lever
15. Snap ring
16. Gasket

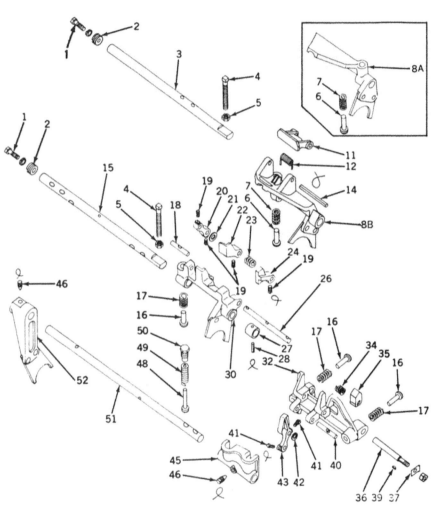

Fig. JD1885—Exploded view of 720 & 730 transmission shifter shafts and shifters.

1. Cap screw
2. Adjusting screws
3. Overdrive shifter shaft
4. Set screw
5. Jam nut
6. Shifter pawl
7. Shifter pawl spring
8A. Overdrive shifter (prior 7202781)
8B. Overdrive shifter (after 7202780)
11. Overdrive shifter lock (after 7202780)
12. Spring (after 7202780)
14. Roll pin (after 7202780)
15. Shifter shaft
16. Shifter pawl
17. Shifter pawl spring
18. Shifter pawl lock
19. Set screw
20. Fourth and sixth speed shifter link
21. Washer
22. Sixth speed shifter arm
23. Spring
24. Fourth speed shifter arm
26. Fourth and sixth speed shifter arm shaft
27. Fourth and sixth speed shifter spacer
28. Roll pin
30. Fourth and sixth speed shifter
32. Second, fifth and reverse speed shifter
34. Spring
35. Second speed shifter arm
36. Second and fifth speed shifter arm shaft
37. Lock plate
39. Woodruff key
40. Shifter pawl lock
41. Set screw
42. Washer
43. Fifth speed shifter arm
45. First and third speed shifter
46. Set screw
48. Shifter pawl
49. Spring
50. Pawl retainer
51. First and third speed shifter shaft
52. First and third speed shifter arm

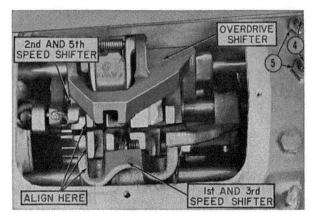

Fig. JD1885A — Series 720 & 730 shifters installation, showing proper alignment of shifter gates.

the raised position by inserting a cotter pin in the pawl hole. On models after Ser. No. 7202780, raise shifter lock (11), slide overdrive shifter (8B) along shaft (3) until the detent pawl rises and hold the pawl in the raised position by inserting a cotter pin in the pawl hole. On all models, loosen both jam nuts (5) and back-off both set screws (4). Remove cap screw (1) and adjusting screw (2) from left end of shaft (3), then turn shaft (3) until detents are out of alignment, withdraw shaft (3) from left side of main case and remove the overdrive shifter (8A or 8B) from above.

Release the pawl locks, slide the fourth and sixth speed shifter (30) and the second, fifth and reverse shifter (32) along shaft (15) until the detent pawls rise and hold the pawls in the raised position by inserting a cotter pin in the pawl holes. Remove cap screw (1) and adjusting screw (2) from left end of shaft (15). Unwire and remove roll pin (28) which positions spacer (27) on the shaft. Turn shaft (15) until detents are out of alignment, withdraw shaft (15) from left side of main case and remove shifter assemblies from above.

Remove the threaded plug (50), then unwire and remove set screw (46) from shifter (45). Withdraw shifter fork (52) and shaft (51) from left and remove shifter (45) from above.

New shifter yokes can be riveted to shifters if old yokes are worn or bent. Renew any shifter shaft that is worn around the detents. Renew any pawl that is worn out-of-round at ball end.

When reinstalling the shifter shafts and shifters, refer to Fig. JD1885 as a guide and reverse the removal procedure. Position all shifters in neutral and turn adjusting screw (2) at left end of center shaft until the gate in the shifters align with the gate in the first and third speed shifter as shown in Fig. JD1885A. But be sure that the center shifter shaft is held to the left against the adjusting screw when

making the adjustment. Tighten set screw (4) and jam nut (5), at right end of shaft, securely; then install and tighten cap screw (1— Fig. JD1885) in left end of shaft. Turn adjusting screw (2) at left end of front shaft until the

left edge of the overdrive shifter aligns with left edge of lug on second and fifth speed shifter as shown in Fig. JD1885A. But be sure that the front shifter shaft is held to the left against the adjusting screw when making the adjustment. Tighten set screw (4) and jam nut (5), at right end of shaft, securely; then install and tighten cap screw (1—Fig. JD1885) in left end of shaft.

139. **SLIDING GEAR SHAFT.** To remove the sliding gear shaft (57— Fig. JD1886) and gears, remove the transmission case top cover as outlined in paragraph 137 and the shifter shafts and shifters as in the preceding paragraph 138. Remove the sheet metal oil retainer (61) and extract outer snap ring (58) from main case. Withdraw the sliding gear shaft from left side of main case and remove

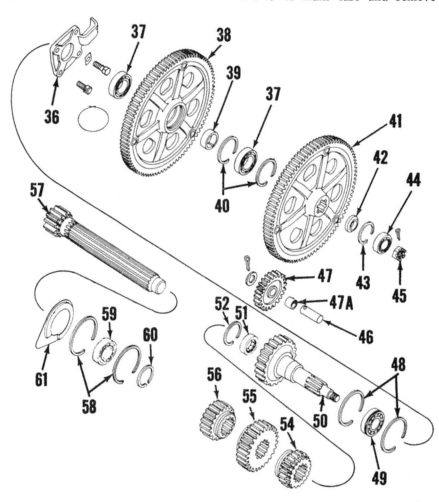

Fig. JD1886—Exploded view of 720 & 730 transmission sliding gear shaft and associated parts. Oil slinger gear (47) is mounted in the first reduction gear cover as shown in Fig. JD1883.

36. Bearing cover	46. Oil slinger gear pin	drive gear
37. Bearing	47. Oil slinger gear	55. Fourth and sixth
38. Powershaft idler	47A. Bushing	sliding pinion
39. Spacer	48. Snap ring	56. Second, fifth and
40. Snap ring	49. Bearing	reverse sliding pinion
41. First reduction gear	50. Transmission drive	57. Sliding gear shaft
42. Spacer	shaft	58. Snap ring
43. Snap ring	51. Pilot bearing	59. Bearing
44. Bearing	52. Snap ring	60. Snap ring
45. Nut	54. Sliding gear shaft	61. Oil retainer

gears from above. Pilot bearing (51) can be removed from the transmission drive shaft (50) after removing snap ring (52).

Using Fig. JD1886 as a guide, reinstall the sliding gear shaft by reversing the removal procedure. Install outer snap ring (58) with gap in snap ring spanning the oil passage in main case. Install the sheet metal oil retainer (61) with flat spot adjacent to oil passage in main case.

140. TRANSMISSION DRIVE SHAFT. To remove the transmission drive shaft (50—Fig. JD1886), remove the clutch, belt pulley and the first reduction gear cover as outlined in paragraph 136. Remove the sliding gear shaft as in the preceding paragraph 139. Withdraw the first reduction gear (41) and idler gear (38). Remove the countershaft right bearing housing (71—Fig. JD1887) so the countershaft assembly will drop down enough to clear the gear on inner end of the drive shaft. Remove bearing cover (36—Fig. JD1886) and withdraw drive shaft from tractor.

When reinstalling the shaft, refer to Fig. JD1886 as a reference and reverse the removal procedure.

140A. COUNTERSHAFT. To remove the transmission countershaft (79—Fig. JD1887), remove the sliding gear shaft as in paragraph 139 and the transmission drive shaft as in paragraph 140. Unstake and remove nut (63) from left end of countershaft. On models prior to Ser. No. 7215213 equipped with the two-piece spacer (70), unbolt and remove the spacer halves. On other models, bump countershaft toward right and spacer toward left until snap ring (70A) comes out from under spacer (70B); then disengage the snap ring from the shaft groove. On all models, bump

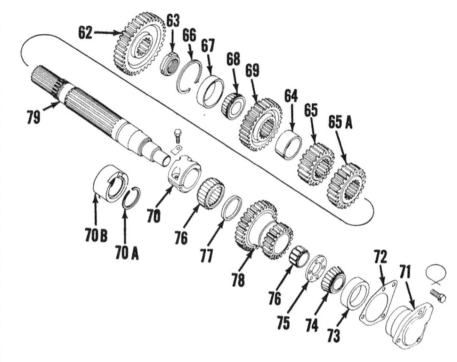

Fig. JD1887—Exploded view of 720 & 730 transmission countershaft. Shaft end play of 0.001-0.004 is controlled by shims and gaskets (72).

62. First and third speed sliding gear	67. Bearing cup	71. Bearing housing
63. Nut	68. Bearing cone	72. Shims and gaskets
64. Spacer	69. Second and fifth speed gear	73. Bearing cup
65. Differential drive pinion	70. Spacer (prior 7215213)	74. Bearing cone
65A. Fourth and sixth speed gear	70A. Snap ring (after 7215212)	75. Thrust washer
66. Snap ring	70B. Spacer (after 7215212)	76. Roller bearing
		77. Spacer
		78. Countershaft idler gear
		79. Countershaft

countershaft out right side of case and remove gears from above.

On early models with the two-piece spacer (70), it is recommended that the two-piece spacer be discarded and the late production one-piece spacer (70B) and snap ring (70A) be installed.

When reinstalling the countershaft, use the same number of shims (72—Fig. JD1887) as were originally removed and tighten nut (63) securely.

Mount a dial indicator with contact button resting on left end of countershaft and check the countershaft end play which should be 0.001-0.004. If end play is not as specified, remove the countershaft right bearing housing and add or remove the necessary amount of shims (72). After the countershaft end play is properly adjusted, remove bearing housing (71); then reinstall same after the transmission drive shaft is in place.

DIFFERENTIAL, FINAL DRIVE AND REAR AXLE

The differential unit is mounted on the spider of a spur (ring) gear which meshes with a driving pinion on the transmission countershaft. Pressed through the spider is the differential cross shaft which forms the journals for the integral differential side gears and spur (bull) pinions. The outer ends of the differential cross shaft carry taper roller bearings which support the differential unit. The remainder of the final drive includes the final drive (bull) gears, located in the rear axle housing, and the wheel axle shafts to which the bull gears are splined.

DIFFERENTIAL
Series 520-530

141. REMOVE AND REINSTALL. To remove the differential, first drain hydraulic system and main case. Remove clutch, belt pulley and both brake assemblies. Remove the first reduction gear cover as in paragraph 123 and withdraw the first reduction gear and the powershaft idler gear. Remove platform and disconnect the hydraulic "Powr-Trol" lines. Remove seat and disconnect battery cable and

wiring harness at rear of tractor. Support tractor under main case and unbolt rear axle housing from main case. With the rear axle housing assembly supported so that it will not tip, roll the unit rearward and away from tractor. Remove both of the differential bearing quills (1 and 14—Fig. JD1890) and withdraw the differential assembly, left hand end first.

When reinstalling, be sure to renew packings (6), use the same thickness of shims and gaskets (2) as were orig-

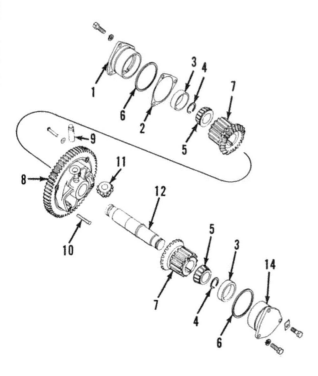

Fig. JD1890 — Exploded view of 520 & 530 differential. The spur ring gear and spider are available as an assembled unit only.

1. Left bearing quill
2. Shims
3. Bearing cup
4. Snap ring
5. Bearing cone
6. Packing
7. Differential side gear and bull pinion
8. Ring gear and spider
9. Pinion shaft
10. Rivet
11. Bevel pinion
12. Differential cross shaft
14. Right bearing quill

Fig. JD1893 — When removing 620, 630, 720 & 730 differential, withdraw the right end first. If the differential only is to be removed, it is not necessary to remove both brake assemblies. Refer to text.

inally removed, mount a dial indicator so that contact button is resting on side of the ring gear and check the differential end play which should be 0.001-0.004. If end play is not as specified, remove the left bearing quill and add or remove the required amount of shims and gaskets (2).

142. **OVERHAUL.** With the differential removed as outlined in the preceding paragraph, proceed to disassemble the unit as follows: Remove snap rings (4) and using a suitable puller, remove both of the differential bearing cones (5) and withdraw the combination differential side gears and spur (bull) pinions (7). Remove rivets (10), extract pinion shafts (9) and remove bevel pinions (11).

Neither the differential spider or spur ring gear (8) is available separately. If either part is damaged, renew the entire assembly. Check the disassembled parts against the values which follow:

I.D. of pinions (7)......2.069-2.071

Diameter of differential
shaft (12) 2.0645-2.0655

NOTE: If shaft (12) is worn, press the old shaft out, support differential spider on a piece of pipe and press the new shaft in place as shown in Fig. JD1891.

I.D. of bevel pinions
(11—Fig. JD1890) 0.863-0.865

Diameter of pinion
shafts (9) 0.8585-0.860

When reassembling, reverse the disassembly procedure and press bearing cones (5) on the differential shaft until they seat. Install new snap rings (4).

Fig. JD1891—When pressing the differential cross shaft in differential, be sure to support the spider with a piece of pipe as shown. The differential shown is a model 50, but the same procedure applies to other models.

Fig. JD1892 — Exploded view of 620, 630, 720 & 730 (except Hi-Crop) differential assembly. Differential end play is adjusted with shims (2).

1. Left bearing quill
2. Shims
3. Bearing cup
4. Snap rings
5. Bearing cone
7. Differential side gear and bull pinion
8. Ring gear and spider
9. Pinion shaft
10. Rivet
11. Bevel pinion
12. Differential cross shaft
15. Snap ring
16. Bearing cover

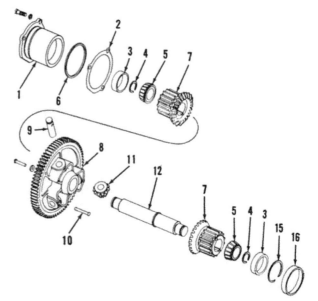

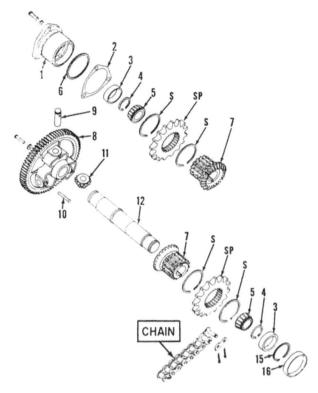

Fig. JD1894 — Exploded view of 620, 630, 720 & 730 "Hi-Crop" differential, drive sprockets and chains. Sprockets (SP) are located on the differential side gears by snap rings (S). Refer to legend under Fig. JD1892.

Series 620-630-720-730

143. **REMOVE AND REINSTALL.** To remove the differential, first drain the hydraulic system and main case. Remove the platform and disconnect the hydraulic "Powr-Trol" lines. Remove seat and disconnect battery cable and wiring harness at rear of tractor. On "hi-crop" models, remove the drive chains as in paragraph 148A. Support tractor under main case and unbolt rear axle or drive housing from main case. With the rear axle or drive housing assembly supported so that it will not tip, roll the unit rearward and away from tractor. On all except "hi-crop" models, remove the left brake assembly and on "hi-crop" models, disengage inner snap rings (S—Fig.

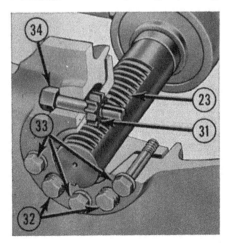

Fig. JD1895—Rear wheels can be removed by loosening cap screws (33), turning jack screws (32) in and turning pinion screw (34).

JD1894) and slide the snap rings and drive sprockets in toward differential. Remove the differential left bearing quill (1—Figs. JD1892 or JD1894) and withdraw the differential assembly, right end first, as shown in Fig. JD 1893.

When reinstalling, be sure to renew packing (6—Figs. JD1892 or JD1894), use the same thickness of shims (2) as were originally removed, mount a dial indicator so that contact button is resting on side of the ring gear and check the differential end play which should be 0.001-0.004. If end play is not as specified, remove the left bearing quill and add or remove the required amount of shims (2).

144. **OVERHAUL.** With the differential removed as outlined in the preceding paragraph, proceed to disassemble the unit as follows: Remove snap rings (4—Figs. JD1892 or 1894). Use a suitable puller and remove the bearing cones (5). Withdraw the combination differential side gears and spur (bull) pinions (7). Remove rivets (10), extract pinion shafts (9) and remove bevel pinions (11).

Neither the differential spider or spur ring gear (8) is available separately. If either part is damaged, renew the entire assembly. Check the disassembled parts against the values which follow:

I.D. of pinions (7)
Series 620-630 ....2.238-2.240
Series 720-730 ....2.597-2.599

Diameter of differential shaft (12)
Series 620-630 ....2.234-2.235
Series 720-730 ....2.592-2.593
NOTE: If shaft (12) is worn, press the old shaft out, support differential spider on a piece of pipe and press the new shaft in place in a manner similar to that shown in Fig. JD1891.
I.D. of bevel pinions (11—Figs. JD1892 or JD1894)
620-630-720-730 ....1.114-1.116
Diameter of pinion shafts (9)
620-630-720-730 .....1.1085-1.1100
When reassembling, reverse the disassembly procedure and press bearing cones (5) on the differential shaft until they seat. Install new snap rings (4).

FINAL DRIVE
All Models
(Except "Hi-Crop")

145. **R&R REAR WHEEL.** Support rear of tractor, turn wheel until rack in axle shaft is in the up position and unscrew the three cap screws (33—Fig. JD1895) approximately $\frac{5}{16}$-inch. Turn the two jack screws (32) clockwise until outer groove in each screw is flush with outer surface of wheel hub. Turn pinion shaft screw (34) and remove wheel.

When reinstalling, make certain both rear wheels are set the same distance from the centerline of the tractor. A slotted hole in back of battery box indicates tractor centerline.

146. **AXLE SHAFT OUTER FELT SEAL RENEW.** To renew the rear wheel axle shaft outer felt seal (21—Fig. JD1897), remove wheel as in the preceding paragraph and using a cold chisel and hammer as shown in Fig. JD1896, drive the felt seal retainer (22—Fig. JD1897) from the axle hous-

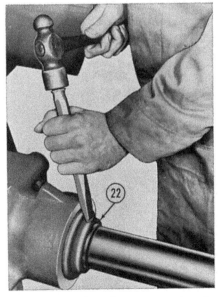

Fig. JD1896 — Removing outer retainer for axle shaft outer felt seal.

ing. Withdraw the seal retainer and felt seal.

When reassembling, renew the seal retainer, and using a brass drift and hammer, drive the retainer in until it seats against recess in housing.

147. FINAL DRIVE (BULL) GEAR, WHEEL AXLE SHAFT & BEARINGS AND/OR AXLE SHAFT INNER OIL SEAL. Drain hydraulic system, transmission and final drive housing. Remove tractor seat and if three-point hitch is installed, remove the upper links, draft links and lift links. Disconnect the draft link supports from the drawbar and loosen the supports attached to rockshaft (basic) housing. Remove the platform and disconnect the hydraulic lines. Disconnect battery cable and wiring harness at rear of tractor. Support the complete basic housing assembly and attach units in a chain hoist so arranged that the complete assembly will not tip. Unbolt basic housing from rear axle housing and move the complete assembly away from tractor.

CAUTION: This complete assembly is heavy and due to the weight concentration at the top, extra care should be exercised when swinging the assembly in a hoist.

Remove cotter pin and loosen, but do not remove the adjusting nut (29—Fig. JD1897). Using a hammer and a long taper wedge as shown in Fig. JD1898, force axle shaft loose from the bull gear. Remove nut (29) and withdraw gear.

Withdraw axle shaft and inspect housing between inner and outer seals for presence of transmission oil. If oil is found, the inner seal (25—Fig. JD1897) should be renewed. The need for further disassembly is evident.

When reassembling, pack the axle shaft outer bearing with wheel bearing grease. Alternately tighten nut (29) and bump outer end of axle shaft to assure proper seating of the taper roller bearings. The proper adjustment is when the axle shaft has an end play of 0.001-0.004; then tighten to the nearest castellation and install

Fig. JD1898—On all models, the axle shaft can be forced out of the final drive (bull) gears by loosening the adjusting nut and driving a long tapered wedge between the nuts.

the cotter pin. When installing the basic housing, turn the power (PTO) shaft to engage the coupling splines.

FINAL DRIVE
Series 620-630-720-730 "Hi-Crop"

148. BULL GEAR, WHEEL AXLE SHAFT, BEARINGS AND/OR SEALS. Support rear of tractor and remove wheel, inner bearing cover (1—Fig. JD1899) and housing lower cover (18). Remove nuts (3), bearing cone (4) and snap ring (21) from inner end of wheel axle shaft (15). Bump the axle shaft out of housing and withdraw the bull gear. The need and procedure for further disassembly is evident after an examination of the unit.

When reassembling, tighten the bearing adjusting nut (3) enough to provide 0.002 deflection of the housing when measured at the inner bearing boss.

148A. R&R CHAIN AND SPROCKET. To remove the drive chains and sprockets, first remove the basic housing or rear cover from the final drive shaft housing. Support rear of tractor and turn the rear wheels until the chain master link is accessible. Remove the master link and withdraw the chains. Remove the nut and washer from inner end of shaft (35—Fig. JD1900) and remove the driven sprocket.

Reinstall in reverse order and tighten nut (24) to obtain a drive shaft end play of 0.001-0.004. Make certain that the drive chain master link is installed with open end toward center of tractor, and check and adjust the chain slack as in the following paragraph.

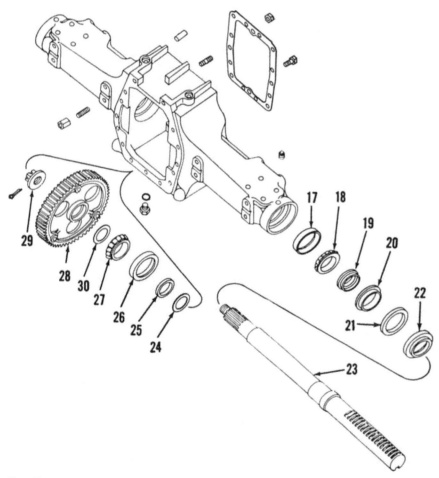

Fig. JD1897—Exploded view of 520 & 530 rear wheel axle shaft and housing assembly. Series 620, 630, 720 and 730 (except "hi-crop") are similar except washer (24) is not used.

17. Bearing cup	21. Felt seal	25. Inner oil seal
18. Bearing cone	22. Outer retainer for felt	26. Bearing cup
19. Bearing spacer	seal	27. Bearing cone
20. Inner retainer for felt	23. Wheel axle shaft	28. Final drive (bull) gear
seal	24. Washer (520 & 530 only)	29. Adjusting nut

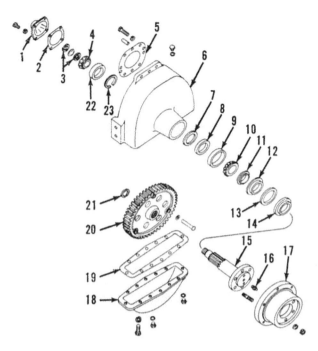

Fig. JD1899 — Exploded view of 620, 630, 720 & 730 "hi-crop" final drive housing, bull gear and wheel axle shaft.

1. Bearing cover
2. Gasket
3. Nut
4. Bearing cone
5. Gasket
6. Housing
7. Washer
8. Oil seal
9. Bearing cup
10. Bearing cone
11. Spacer
12. Inner retainer
13. Felt seal
14. Outer retainer
15. Axle shaft
16. Grease fitting
17. Wheel hub
18. Housing cover
19. Gasket
20. Bull gear
21. Snap ring
22. Bearing cup
23. Snap ring

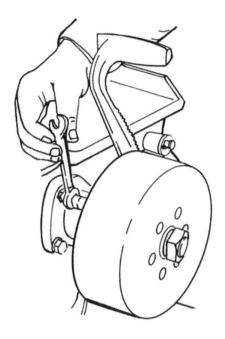

Fig. JD1901—Adjusting the brakes on all models. Brake pedals should have a free travel of approximately three inches.

Note: To remove the drive sprockets which are mounted on the differential side gears, it is necessary to remove the differential as outlined in paragraph 143.

149. ADJUST CHAINS. To check and/or adjust the slack in the drive chains, remove the inspection covers from side of main case. The chains should have ½ to 1¾-inch slack. To adjust the slack, remove the brake drum and the brake housing retaining bolts. Turn the brake housing one hole at a time, until the desired chain slack is obtained and reinstall the housing retaining bolts.

After adjusting the chain slack, remove the brake shoe carrier from the brake housing and re-position the carrier so that brake pedal is in the proper position.

149A. R&R DRIVE SHAFT AND PINION. To remove the drive shaft (35—Fig. JD1900) support rear of tractor, remove the driven sprocket (26) as outlined in paragraph 148A and remove the final drive (bull) gear housing from the drive shaft housing. Withdraw the drive shaft from the housing.

When reassembling, tighten nut (24) enough to provide the drive shaft with an end play of 0.001-0.004. Tighten the drop housing to drive shaft housing screws to a torque of 275 Ft.-Lbs.

BRAKES

All Models (Except "Hi-Crop")

150. ADJUSTMENT. To adjust the brakes, tighten the adjusting screw as shown in Fig. JD1901 to reduce the pedal free travel to approximately three inches.

151. R&R BRAKE SHOES. To remove the brake shoes for lining replacement, loosen adjusting screw (41 —Fig. JD1902), remove nut (56) and using a rawhide hammer, bump brake shaft (37) inward to free drum from shaft. Withdraw the brake drum, pry shoes away from adjusting pins to release spring tension and remove shoes.

Install brake shoes by reversing the removal procedure and adjust the brakes as outlined in paragraph 150.

152. R&R AND OVERHAUL BRAKE ASSEMBLY. On 520, 530, 620 and 630 tractors, use a double nut arrangement and remove the upper inside implement mounting stud from the front face of rear axle housing before attempting to unbolt and remove the brake assembly. The removal procedure is evident on 720 and 730 tractors.

The procedure for disassembling the brakes is evident after an examination of the unit and reference to Fig. JD-1902. Be sure to mark the relative position of pedal shaft (42) with re-

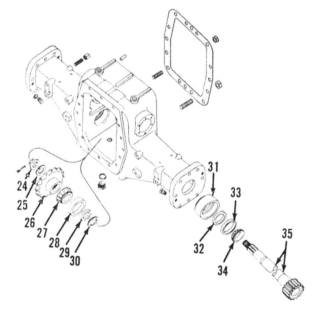

Fig. JD1900 — 620, 630, 720 & 730 "hi-crop" final drive shaft and housing. Nut (24) should be tightened enough to provide shaft (35) with an end play of 0.001-0.004.

24. Nut
25. Washer
26. Driven sprocket
27. Bearing cone
28. Bearing cup
29. Oil seal
30. Washer
31. Adapter
32. Oil seal
33. Bearing cup
34. Bearing cone
35. Drive shaft

spect to pedal (39) and housing (44). Check the disassembled parts against the values which follow:

I.D. of inner bushing (38)
Series 520-530 ....1.6215 Min.
Series 620-630 ....1.7520 Min.
Series 720-730 ....2.0030 Min.

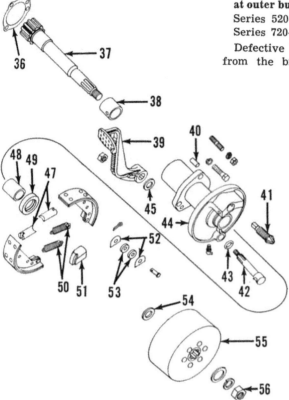

Diameter of brake shaft at inner bushing
Series 520-530 ....1.619-1.620
Series 620-630 ....1.749-1.750
Series 720-730 ....1.999-2.000

I.D. of outer bushing (48)
Series 520-530-620-630...1.4965 Min.
Series 720-730 ....1.5000 Min.

Diameter of brake shaft at outer bushing
Series 520-530-620-630...1.494-1.495
Series 720-730 ....1.494-1.495

Defective bushings can be removed from the brake housing by sawing

Fig. JD1902 — Exploded view of typical brake used on all except "hi-clearance" models. Adjustment is accomplished by turning screw (41).

36. Gasket
37. Brake shaft
38. Inner bushing
39. Pedal
40. Dowel pin
41. Adjusting screw
42. Pedal shaft
43. "O" ring
44. Housing
45. Washer
47. Adjusting pin
48. Outer bushing
49. Oil seal
50. Spring
51. Cam
52. Washer
53. Rollers
54. Thrust washer
55. Brake drum
56. Nut

through the bushing wall with a hack saw and driving bushing out, but be careful not to damage the brake housing during the sawing operation.

Install new bushings by using a suitable piloted drift and to a depth of $\frac{1}{16}$-inch as shown in Fig. JD1903.

When reassembling, vary the number of thrust washers (54—Fig. JD-1902) to provide a brake shaft end play of 0.004-0.044.

Series 620-630-720-730 "Hi-Crop"

153. **ADJUSTMENT.** The procedure for adjusting the brakes is the same as outlined in paragraph 150.

154. **R&R BRAKE SHOES.** To remove the brake shoes for lining replacement, loosen adjusting screw (41—Fig. JD1904), remove the brake drum guard, remove nut (56) and remove drum. Pry the shoes away from the adjusting pins to release spring tension and remove shoes.

Install shoes by reversing the removal procedure and adjust the brakes as in paragraph 150.

155. **R&R AND OVERHAUL BRAKE ASSEMBLY.** To remove the brake assembly, first remove the transmission rear cover or basic housing, extract snap ring (57—Fig. JD1904) and remove sprocket (58). Remove the brake drum guard, then unbolt and remove assembly from tractor. Before disassembling the unit, mark the relative position of pedal shaft (42) with respect to pedal (39) and carrier (59).

The remainder of the overhaul procedure is similar to the brakes used on non-"hi-crop" models. Refer to paragraph 152. When reinstalling the brakes, adjust the drive chain slack as in paragraph 149.

Fig. JD1903 — The brake shaft bushings should be installed to a depth of 1/16-inch below end of housing as shown.

Fig. JD1904 — Exploded view of brakes used on "Hi-Crop" models.

36. Gasket
37. Brake shaft
38. Inner bushing
39. Pedal
41. Adjusting screw
42. Brake lever shaft
43. "O" ring
44. Housing
45. Washer
47. Adjusting pin
48. Outer bushing
49. Dust guard
50. Spring
51. Brake cam
52. Washer
53. Roller
54. Washer
55. Drum
56. Nut
57. Snap ring
58. Sprocket
59. Brake shoe carrier

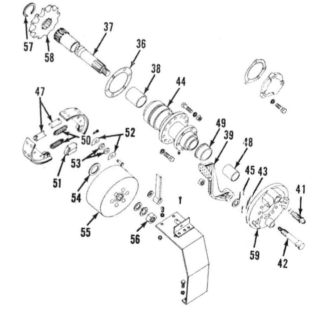

POWER TAKE-OFF SYSTEM

CLUTCH, OUTPUT SHAFT, DRUM AND OIL PUMP

All Models

165. ADJUST CLUTCH. To adjust the clutch on 530, 630, 730, 520 after Ser. No. 5206491, 620 after Ser. No. 6210213, 720 after Ser. No. 7210335, and earlier models on which the improved powershaft clutch has been installed, proceed as follows: Remove the cap nut, exposing the adjustment indicator rod (21—Fig. JD1905) and guide (20). Engage the clutch, then pull back on the clutch pedal to take up all slack in the linkage but do not disengage the clutch. Hold the adjustment indicator rod in and note the position of the rod with respect to end of rod guide (20). Disengage clutch and be sure pedal is latched, then again note position of indicator rod (21) in relation to end of guide. If clutch is properly adjusted, the distance the rod has moved will be equal to the length of one land and one groove (0.090) on the rod as shown. If adjustment is not as specified, proceed as follows:

If tractor is equipped with a universal three-point hitch, remove the right hand draft link and draft link support. Remove the clutch adjusting hole cover from side of clutch housing, and with the clutch pedal in neutral (disengaged but not latched), turn the power (output) shaft until the adjusting cam locking screw is visible through opening in clutch housing as shown in Fig. JD1906. Now, latch the pedal in the disengaged position and turn the locking screw in until head

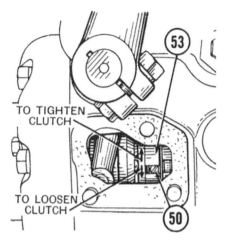

Fig. JD1906 — Side view of powershaft clutch housing with adjusting hole cover removed. Adjusting cam (53) is locked in position with screw (50).

of same clears slot in the adjusting cam. To tighten the clutch, turn the adjusting cam counter-clockwise (viewed from rear), one notch at a time and recheck the adjustment. Continue this procedure until adjustment is as specified, then turn the locking screw outward into one of the notches of the adjusting cam. Reinstall the adjusting hole cover and indicator rod cap nut.

If the powershaft fails to stop when clutch is disengaged and pedal latched, adjust the powershaft brake as follows: Disconnect yoke (Fig. JD1907) at rear end of operating rod and shorten the rod ½-turn at a time until proper adjustment is obtained.

The procedure for adjusting the powershaft clutch on early models on which the improved powershaft clutch

has not been installed, is similar to the above procedure except there is no indicator rod for checking the adjustment. Proper adjustment is determined by measuring how far the clutch cam disc moves rearward from the brakeplate as the clutch is engaged. The desired movement of 0.090 can be checked by using a thin ruler through plug hole in clutch housing as shown in Fig. JD1908 or, by using a $\frac{3}{32}$-inch wire through opening in side of clutch housing as shown in Fig. JD1909. Keep in mind, however, that power (output) shaft on these early models can be turned only when the clutch is engaged.

167. OVERHAUL CLUTCH AND OUTPUT SHAFT. To remove the clutch and/or power (output) shaft, first drain oil from clutch housing and engage the clutch. Remove the

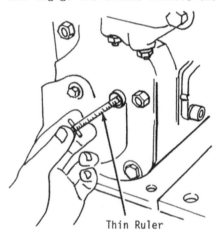

Thin Ruler

Fig. JD1908—Using a thin ruler to check the powershaft clutch adjustment on early 520, 620 and 720 models.

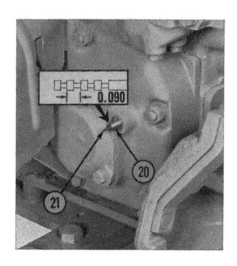

Fig. JD1905—Checking powershaft clutch adjustment. Refer to text for procedure.

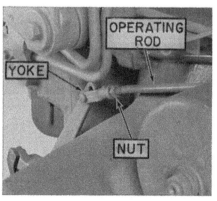

Fig. JD1907—The length of the powershaft clutch operating rod can be varied to obtain proper adjustment of the powershaft brake.

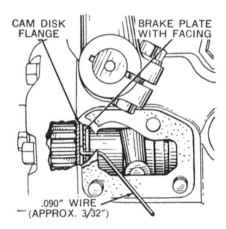

CAM DISK FLANGE   BRAKE PLATE WITH FACING

.090" WIRE (APPROX. 3/32")

Fig. JD1909 — Using 3/32-inch rod to check powershaft clutch adjustment on early 520, 620 and 720 models.

powershaft guard and disconnect the clutch operating linkage. Remove the cap screws and nuts retaining the clutch housing cover to the clutch housing and withdraw clutch, cover and powershaft assembly. It may be necessary to pry cover from its locating dowels.

NOTE: The clutch drum cannot be serviced at this time. If the drum requires renewal, refer to paragraph 168.

Inspect the clutch fork (25—Fig. JD1910) and fork shoes (23). To remove the fork, punch assembly marks on the fork and shaft, remove ring (26) and withdraw fork shaft (27).

167A. Support the removed clutch and housing cover assembly in a soft jawed vise and using two large screw drivers, disengage the clutch as shown in Fig. JD1911. Remove snap ring (43) retaining clutch plates to clutch shaft and if there are shim washers under the snap ring, save them for reinstallation. Withdraw clutch plates, discs, release springs, adjusting cam, balls and clutch collar. On early models so equipped, unlock the brake plate retaining cap screws, unscrew the screws evenly to avoid damaging the brake plate and withdraw cam with disc, springs and washer. If there are shims under the brake plate, save them for reinstallation. On later models, the brake plate is retained by snap rings on the support pins.

Remove bearing cover (34—Fig. JD-1912), and oil seal housing (55). Extract snap ring (9) from rear end of clutch shaft and bump or press clutch shaft (11) out of bearing (10). Lift powershaft and gear assembly from cover.

To disassemble the clutch shaft on late models, press the clutch cam down for enough to remove snap ring (61).

Lift off the hardened washer (63) and adjusting washer (62).

The nine outer clutch pressure springs (14) should require 83-93 pounds to compress them to a height of $1\frac{5}{8}$ inches.

The clutch release springs (48) should require $12\frac{1}{2}$-$15\frac{1}{2}$ pounds to compress them to a height of $2\frac{9}{16}$ inches on 720 and 730, $1\frac{21}{32}$ inches on 620 and 630, $1\frac{9}{16}$ inches on 520 and 530.

Inspect clutch shaft and pilot bushing (41) which is located in clutch drum. If new bushing is carefully installed with a piloted mandrel, it will not require final sizing.

Inspect clutch facings to make certain they are in good condition and be sure plates and discs are not worn or out-of-flat more than 0.008. Inspect all other parts and renew any which are questionable.

On late models, assemble the nine large springs (14) in cam (15), then install a small spring (13) into each third large spring. Lay washer (12) over the springs and install shaft (11). Place the assembly on a press as shown in Fig. JD1913 and press the cam down until springs are compressed solid. Then release pressure on the cam and allow it to move upward $\frac{5}{32}$-inch only. Install sufficient adjusting washers (62) under the hardened cam washer so snap ring (61—Fig. JD1912) can just be inserted. Install snap ring, but make certain that the hardened washer is next to the snap ring. Install clutch shaft bearing into cover so that shielded side of bearing is toward inside, then place powershaft assembly in cover. Install bearing washer (W), press clutch shaft assembly into position and install snap ring (9). Install brake plate (19) with facing side next to flange on cam then install retainer rings on the support pins and clamp

the rings tightly into the pin grooves. Install clutch shaft bearing cover (34).

On early models which do not have the improved clutch, install the clutch shaft bearing (10) into cover, press clutch shaft into position and install snap ring (9). Place powershaft assembly into cover. Install washer (12) and place the nine large springs in position on washer (12), then install a small spring (13) into each third large spring. Place cam (15) over the springs. If there were shims (S) installed between the brake plate and housing, install the same quantity as were removed. Install brake plate (19) with facing side next to cam and turn the cap screws (SS) down evenly, a little at a time, to avoid bending or breaking the brake plate. When the cap screws are tight, lock them in position with the tab washers. Install clutch shaft bearing cover (34).

On all models, install clutch collar (22), balls (17) and adjusting cam assembly (49, 50, 51 and 52). Turn the adjusting cam until hub (49) is about $\frac{1}{16}$-inch out of the cam. Install plate (45) with facing up, followed by a steel drive disc (46). Install a faced disc (47) with opening for release spring in line with oil hole in shaft, then install another steel drive disc (46). Continue alternating the discs in this order until all lined and unlined plates are installed. Install springs (48). Install the thick driven plate (45) with facing down and install snap ring (43). NOTE: If washers (44) were originally installed under the snap ring, be sure the same number are in place. Install oil seal (54) in oil seal housing so that lip of seal will face front of tractor then install the oil seal housing and tighten the screws finger tight.

Fig. JD1910 — With the powershaft clutch removed, the clutch fork (25) can be removed after removing the retainer ring.

Fig. JD1911—Disengaging the pto clutch, using two large screw drivers. Removal of snap ring (43) will permit disassembly of the clutch.

To adjust the clutch on 530, 630, 730, 520 after Ser. No. 5206491, 620 after Ser. No. 6210213, 720 after Ser. No. 7210335, and earlier models on which the improved powershaft clutch has been installed, proceed as follows: Remove the cap nut, exposing the adjustment indicator rod (21—Fig. JD-1905) and guide (20) and note the position of the rod with respect to end of rod guide (20) when clutch is disengaged. Engage clutch and again note the position of indicator rod (21) in relation to end of guide. If clutch is properly adjusted, the distance the rod has moved will be equal to the length of one land and one groove (0.090) on the rod as shown. If indicator rod movement is insufficient, disengage clutch, using two pry bars

as shown in Fig. JD1911, turn the cam locking screw inward until head of same clears slot in adjusting cam and turn the adjusting cam counter-clockwise (viewed from rear) to increase the indicator rod movement, then recheck the adjustment. If indicator rod moves an excessive amount, turn the adjusting cam in the opposite direction.

When the clutch is properly adjusted, turn the locking screw outward into one of the notches in the adjusting cam.

After the clutch is properly adjusted, disengage it and check the clearance between adjusting cam and rear heavy clutch plate as shown in Fig. JD1914. The adjusting cam should not contact the clutch plate and at

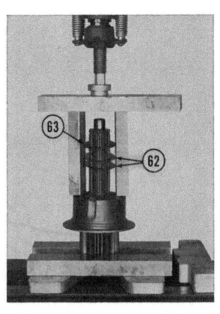

Fig. JD1913—Compressing cam for the purpose of installing adjusting washers (62), hardened washer (63) and snap ring.

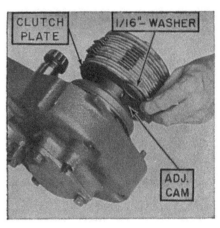

Fig. JD1914 — Using 1/16-inch washer to check adjusting cam clearance.

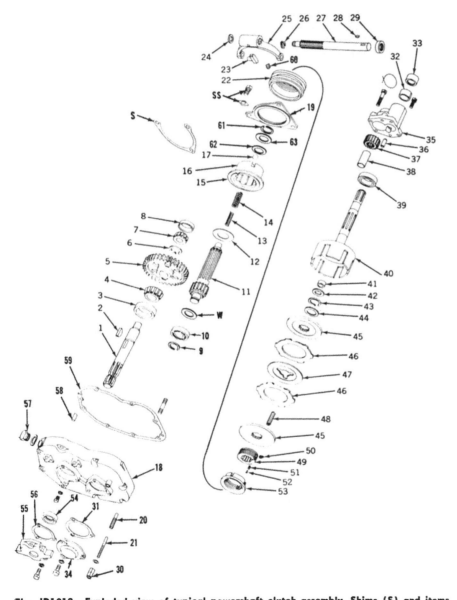

Fig. JD1912—Exploded view of typical powershaft clutch assembly. Shims (S) and items (SS) are used only on early models which have not been converted to the later construction. Washer (W) and items (61, 62 and 63) are used only on late models and early models which have been converted to the late construction.

1. Power (output) shaft	34. Bearing cover
3. Bearing cup	35. Pump housing
4. Bearing cone	36. Dowel pin
5. Powershaft gear	37. Idler gear
6. Snap ring	38. Idler gear shaft
7. Bearing cone	39. Bearing
8. Bearing cup	40. Clutch drum and shaft
9. Snap ring	41. Bushing
10. Bearing	42. Thrust washer
11. Clutch shaft	43. Snap ring
12. Washer	44. Washer
13. Spring	45. Clutch plate with facing
14. Spring	46. Clutch drive disc
15. Clutch cam disc	47. Clutch disc with facing
16. Clutch cam	48. Spring
17. Ball	49. Clutch adjusting hub
18. Clutch cover	50. Set screw
19. Brake plate with facing	51. Spring
20. Adjustment indicator guide	52. Rivet
21. Adjustment indicator	53. Clutch adjusting cam
22. Clutch collar	54. Oil seal
23. Clutch fork shoe	55. Oil seal housing
24. Washer	56. Shim
25. Clutch fork	57. Oil filler plug
26. Snap ring	58. Dowel pin
27. Clutch fork shaft	59. Gasket
28. Woodruff key	60. Clutch collar pin (520 only)
29. Oil seal	61. Snap ring
30. Cap nut	62. Washer
31. Gasket	63. Washer
32. Bushing	
33. Oil seal	

Fig. JD1915—Using feeler gage to check early model pto clutch adjustment prior to final assembly on tractor.

the same time, the clearance should not exceed $\frac{1}{16}$-inch. If a $\frac{1}{16}$-inch washer can be inserted, loosen the clutch adjustment and add an adjusting washer (44—Fig. JD1912). If the adjusting cam contacts the clutch plate, remove an adjusting washer (44) and recheck the clutch adjustment.

The procedure for adjusting the powershaft clutch on early models on which the improved clutch has not been installed, is similar to the above procedure except there is no indicator rod for checking the adjustment. Proper adjustment is determined by measuring how far the clutch cam disc moves rearward from the brake plate as the clutch is engaged. The desired movement of 0.090 can be checked with a feeler gage as shown in Fig. JD1915.

Before installing the clutch, use a spare clutch drum or straight edge, align clutch plates and engage clutch. Install a new thrust washer (42—Fig. JD1912) with smooth side toward shaft, install the complete assembly, making certain that clutch fork shoes engage grooves in clutch collar. Tighten the clutch cover retaining screws to a torque of 56 Ft.-Lbs.

Mount a dial indicator in a suitable manner and check the power (output) shaft end play which should be 0.001-0.004. If end play is not as specified, vary the number of gaskets (56) and recheck.

168. OVERHAUL CLUTCH DRUM AND OIL PUMP. To remove the clutch drum, first drain hydraulic system, clutch housing and main case. Remove clutch, clutch housing cover assembly and clutch fork as outlined in paragraph 167. Remove platform, disconnect hydraulic lines and remove seat. Disconnect battery cable and wiring harness at rear of tractor. Support the complete basic housing assembly and attached units in a chain hoist so arranged that the complete assembly will not tip. Unbolt basic housing from rear axle housing and move the complete assembly away from tractor.

CAUTION: This complete assembly is heavy and due to the weight concentration at the top, extra care should be exercised when swinging the assembly in a hoist.

Remove the cap screws retaining the clutch oil pump body to front side of basic housing and remove pump body and idler gear as shown in Fig. JD1916. Pull clutch drum (40) from basic housing.

The drum shaft is supported in an anti-friction bearing (39—Fig. JD-1912) which should be renewed if its condition is questionable.

The front end of the clutch drum shaft is supported by bushing (32). Bushing is pre-sized and will not require reaming if carefully installed.

Check the clutch oil pump parts against the values which follow:

O.D. of idler gear.......2.115-2.117

I.D. of idler gear bore
  in housing 2.121-2.123

O.D. of idler gear shaft .0.9994-1.0000

I.D. of idler gear bushing .1.002-1.003

Thickness of idler gear..0.997-0.998

Depth of idler gear bore
  in housing 1.000-1.005

When reassembling, coat mating surfaces of pump body and basic housing with shellac and use shim stock around splines of drum shaft to avoid damaging the pump housing oil seal. Refer to Fig. JD1917.

DRIVING GEARS, BEVEL GEARS AND SHIFTER
All Models

169. RENEW DRIVE AND IDLER GEARS. To renew the powershaft drive gear (2—Figs. JD1920, JD1921 or JD1922) and/or idler gear (6), first remove engine clutch, belt pulley and first reduction gear cover as outlined in paragraph 123 for 520 and 530, para-

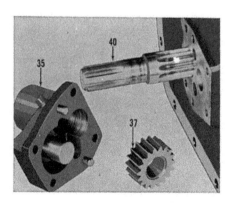

Fig. JD1916—Power shaft clutch oil pump body and idler gear removed from front face of basic (rockshaft) housing. The pump drive gear and shaft is also the clutch drum shaft.

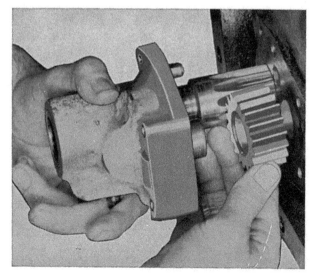

Fig. JD1917—Assembling the power take-off clutch oil pump. Mating surfaces of pump housing and basic housing should be coated with shellac.

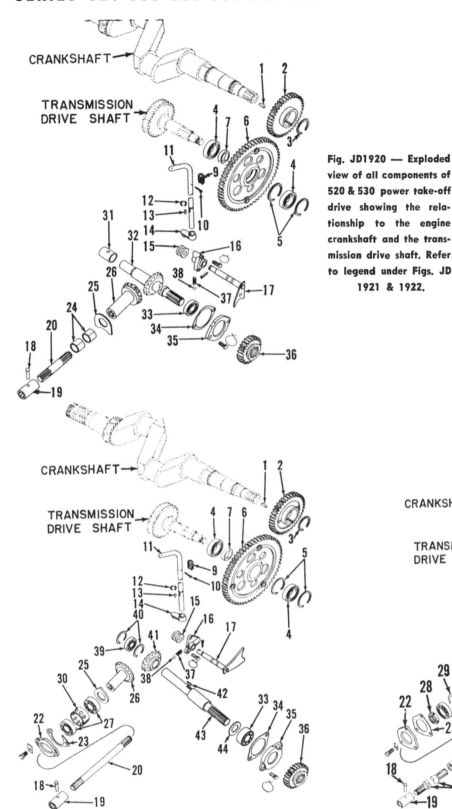

graph 130 for 620 and 630 or paragraph 136 for 720 and 730. Withdraw spacer, first reduction gear and powershaft idler gear. Remove snap ring (3) and using a suitable puller, remove the drive gear from crankshaft.

The idler gear is carried on two anti-friction type bearings (4) which should be inspected and renewed if their condition is questionable.

When reassembling, heat the powershaft drive gear in hot oil and drive it in place with a brass drift.

170. **OVERHAUL 520, 530, 620 & 630 BEVEL GEARS & SHIFTER.** To overhaul the power take-off driving bevel gears and associated parts, first drain the hydraulic system and main case. Remove the platform and disconnect the hydraulic "Powr-Trol" lines. Remove seat and disconnect battery cable and wiring harness at rear of tractor. On "hi-crop" models, remove the drive chains as in paragraph 148A. Support tractor under main case and unbolt rear axle housing from main case. With the rear axle housing assembly supported so that it will not tip, roll the unit rearward and away from tractor.

Fig. JD1920 — Exploded view of all components of 520 & 530 power take-off drive showing the relationship to the engine crankshaft and the transmission drive shaft. Refer to legend under Figs. JD 1921 & 1922.

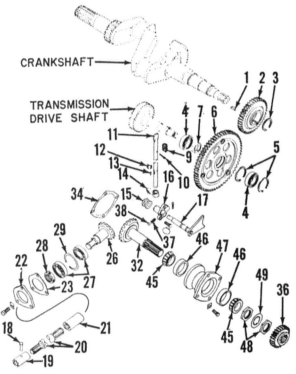

Fig. JD1921—Exploded view of all components of 620 & 630 power take-off drive. The drive train originates with gear (2) which is keyed to the engine crankshaft.

Fig. JD1922—Exploded view of all components of 720 & 730 power take-off drive. Idler gear (6) rotates on the transmission drive shaft.

1. Woodruff key	12. Snap ring	22. Bearing cover (620, 630, 720 & 730)
2. Drive gear	13. Woodruff key	23. Shims (620, 630, 720 & 730)
3. Snap ring	14. Shifter arm	24. Bushings (520 & 530)
4. Bearing	15. Shifter shaft spool	25. Thrust plate (520, 530, 620 & 630)
5. Snap ring	16. Shifter shaft bearing	26. Bevel driven gear
6. Idler gear	17. Shifter shaft and yoke	27. Bearings (620, 630, 720 & 730)
7. Spacer	18. Rivet	
9. Spring	19. Coupling	
10. Cotter pin	20. Coupling shaft	
11. Shifter lever	21. Coupling (720 & 730 only)	

28. Locking nut (720 & 730)	35. Bearing cover (520, 530, 620 & 630)	42. Woodruff key (620 & 630)
29. Snap ring (720 & 730)	36. Sliding gear	43. Drive shaft (620 & 630)
30. Spacer (620 & 630)	37. Spring	44. Washer (620 & 630)
31. Bushing (520 & 530)	38. Ball	45. Bearing cone (720 & 730)
32. Bevel drive gear & shaft (520, 530, 720 & 730)	39. Bearing (620 & 630)	46. Bearing cup (720 & 730)
33. Bearings (520, 530, 620 & 630)	40. Snap ring (620 & 630)	47. Drive shaft quill (720 & 730)
34. Shims	41. Bevel drive gear (620 & 630)	48. Jam nut (720 & 730)
		49. Locking washer

Remove the transmission countershaft as in paragraph 128 for 520 and 530 or 135 for 620 and 630.

Working through top of main case, remove cotter pin (10—Figs. JD1920 or JD1921) from shifter lever and pull the lever part way out of the transmission case. Using a drift, bump shifter arm (14) down and off the shaft. Remove Woodruff key (13), snap ring (12), spring (9) and withdraw shifter lever from above. Refer to Fig. JD1923. Remove cotter pin from inner end of shifter shaft (17— Figs. JD1920 or JD1921), slide spool (15) from shifter shaft and unbolt shifter shaft bearing from main case. Remove shifter shaft (17), bearing (16) and sliding gear (36).

Remove bearing cover (35). bump shaft (32 or 43) toward right until bearing (33) emerges from case and using a suitable puller, remove bearing from shaft. Withdraw drive shaft and pinion through top opening in main case. Reach down through top opening in case and withdraw bevel gear and shaft (26) from main case boss.

On 520 and 530, the drive shaft and bevel gear shaft are carried in bushings in the main case bosses. Check the shafts and bushings against the values which follow:

I.D. of bushings (31).....1.377-1.379

I.D. of bushings (24).....1.632-1.634

O.D. of shaft (26) at
   bushing (24) 1.625-1.626

When installing the drive shaft bushing, make certain that oil hole in bushing is in register with oil hole in bushing boss. When installing the bevel gear shaft bushings, make certain that bushings do not cover oil hole in the bushings boss. The front bushing should extend forward of the bushing boss by 0.082 or approximately three-fourths the thickness of the thrust washer. The rear bushing should be flush with rear of bushing boss. Install bushings with narrow end of oil grooves toward oil hole in boss. Groove in front bushing should be in the 9 o'clock position when viewed from rear and groove in rear bushing should be in the 12 o'clock position when viewed from rear.

On 620 and 630, the drive shaft and bevel gear shaft are carried in bearings as shown in Fig. JD1921. The drive shaft bearing (39) is retained in the case boss by snap rings (40). The bevel gear shaft bearings (27) are retained in the case boss by bearing cover (22).

On all models, when installing the bevel gear and shaft, make certain that smooth side of thrust washer (25) is toward the bevel gear and that the washer fits into the machined recess on front of case boss. On 520 and 530, the thrust washer controls the mesh position of the bevel gears and should not be excessively worn. Thickness of new thrust washer is 0.102-0.106. On 620 and 630, vary the number of shims (23) so that heels of bevel gears are in register. When installing the drive shaft, vary the number of shims (34) between cover (35) and main case to give the bevel gears a backlash of 0.012-0.018.

171. OVERHAUL 720 & 730 BEVEL GEARS & SHIFTER. To overhaul the power take-off driving bevel gears and associated parts, first drain the hydraulic system, main case and first reduction gear cover. Remove the differential as outlined in paragraph 143 and the first reduction gear cover as in paragraph 136. Working through rear opening in main case, remove cotter pin (10—Fig. JD1922) from shifter lever and pull the lever part way out of the main case. Bump shifter arm (14) down and off the shaft. Remove Woodruff key (13), snap ring (12), spring (9) and withdraw shifter lever from above. Remove cotter pin from inner end of shifter shaft (17), slide spool (15) from shifter shaft and unbolt shifter shaft bearing from main case. Remove shifter shaft (17), bearing (16) and sliding gear (36).

Remove cap screws retaining quill (47) to main case and withdraw quill and drive shaft assembly. To disassemble the unit, remove nuts (48). Working through rear opening in main case, remove locking nut (28) and push the driven shaft forward and out of the case boss. Remove bearing retainer (22). The need and procedure for further disassembly is evident.

Refer to Fig. JD1924 and check I.D. of drive shaft quill which should be 1.764-1.766 as shown. This should provide a shaft diametral clearance of 0.016-0.024. If I.D. of quill is too small, ream out the quill to provide the specified clearance.

When installing drive shaft (32 — Fig. JD1922) into quill (47), tighten inner nut (48) enough to provide the shaft with an end play of 0.001-0.005. Install locking washer and outer nut (48). Install the front driven shaft bearing (27) and pull it into case boss until it is against the snap ring. Install rear bearing (27), shims (23) and retainer (22) and tighten the retainer screws finger tight. Install the assembled drive shaft and quill assembly, using the original number of shims (34). When the bevel gears are properly installed, the heels will be in register and the backlash will be 0.012-0.018. If mesh and backlash are not as specified, vary the number of shims (23 and 34) until the proper conditions are obtained.

Install the remaining parts by reversing the removal procedure.

Fig. JD1923—Method of removing the pto shifter lever (11) and shifter arm (33) from main case.

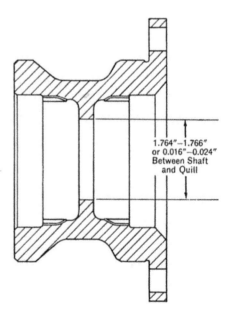

1.764"-1.766" or 0.016"-0.024" Between Shaft and Quill

Fig. JD1924 — On 720 & 730, the inside diameter of the power shaft drive shaft quill should be 1.764-1.766 as shown.

"POWR-TROL" (HYDRAULIC LIFT)

175. Custom "Powr-Trol" is available on general purpose tractors in the following six ways:

1. Rear rockshaft only.
2. Rear rockshaft, load and depth control and three-point hitch.
3. Rear rockshaft and single remote cylinder (with or without load and depth control and three-point hitch).
4. Rear rockshaft and two remote cylinders (with or without load and depth control and three-point hitch).
5. Rear rockshaft, single remote cylinder and single control front mounted rockshaft (with or without load and depth control and three-point hitch).
6. Rear rockshaft, two remote cylinders and double control front mounted rockshaft (with or without load and depth control and three-point hitch).

Custom "Powr-Trol" is available on standard tractors in the following six ways:

1. Rear rockshaft only.
2. Single remote cylinder only.
3. Two remote cylinders only.
4. Rear rockshaft, load and depth control and three-point hitch.
5. Rear rockshaft and single remote cylinder (with or without load and depth control and three-point hitch).
6. Rear rockshaft and two remote cylinders (with or without load and depth control and three-point hitch).

Note: Some models which have the rear rockshaft only, may have a dash pot in the rockshaft housing so they can be converted to load and depth control.

Tractors equipped with a single remote cylinder valve housing will accommodate either a single acting or a double acting remote cylinder; whereas, on tractors equipped with a dual remote cylinder valve housing, the No. 1 circuit will accommodate either a single acting or a double acting remote cylinder, but the No. 2 circuit will accommodate only double acting cylinders.

Note: The maintenance of absolute cleanliness of all parts is of utmost importance in the operation and servicing of the hydraulic system. Of equal importance is the avoidance of nicks or burrs on any of the working parts.

LUBRICATION AND BLEEDING

176. It is recommended that the "Powr-Trol" working fluid be changed twice a year. After the system is completely drained, refill the reservoir, operate the system several times to bleed out any trapped air and refill reservoir to full mark on dip stick.

Note: If other than John Deere cylinders are used, always retard the piston before checking the fluid level.

System capacities are:

520-530, 10 U.S. qts., plus one qt. for each remote cylinder.

620-630, 11 U.S. qts., plus one qt. for each remote cylinder.

720-730, 13 U.S. qts., plus one qt. for each remote cylinder.

For temperatures above 90 degrees F., use SAE 30 single viscosity oil or SAE 10W-30 multi-viscosity oil; for 32 degrees F. to 90 degrees F., use SAE 20-20W single viscosity oil or SAE 10W-30 multi-viscosity oil; for temperatures below 32 degrees F., use SAE 10W single viscosity oil or SAE 5W-20 multi-viscosity oil.

TROUBLE-SHOOTING

177. The following paragraphs should facilitate locating troubles in the hydraulic system.

177A. **INSUFFICIENT LOAD RESPONSE.** Could be caused by:

a. Upper link in low sensitivity hole of load control yoke, paragraph 181.
b. Implement is not level.
c. Implement does not penetrate due to angle of ground engaging elements, check implement operator's manual.
d. Implement not trailing properly, check implement adjustment and alignment of ground engaging elements.
e. Dash pot orifice plugged, paragraph 202.
f. Load control arm to shaft key sheared.
g. Excessive initial loading of load control spring, paragraph 181.
h. Rockshaft control valve opening too wide, paragraph 184.
i. Improper front end weighting, see tractor operator's manual.

177B. **EXCESSIVE LOAD RESPONSE.** Could be caused by:

a. Upper link in high sensitivity hole of load control yoke, paragraph 181.
b. Broken load control arm.
c. Dash pot inlet valve leaking, paragraph 202.
d. Faulty "O" ring on dash pot piston or spring retainer, paragraph 202.
e. Dash pot relief valve leaking, paragraph 187.
f. Dash pot relief valve opening pressure too low, paragraph 187.
g. Low initial loading of load control spring, paragraph 181.
h. Excessive "suck" in implement, see implement operator's manual.
i. Implement ground-engaging elements in poor condition.
j. Implement not trailing properly, check adjustment of implement and alignment of ground engaging elements.

177C. **INSUFFICIENT TRANSPORT CLEARANCE.** Could be caused by:

a. Lift links adjusted too long.
b. Upper link too long.
c. Less than full rockshaft rotation, paragraph 185.
d. Incorrect implement mast angle, see implement operation manual.
e. Implement mast braces too long.
f. Flexible elements of implement not properly supported, see implement operator's manual.
g. Wear or looseness in implement.
h. Implement not level.

177D. **HITCH FAILS TO LIFT.** Could be caused by:

a. Excessive load.
b. Pump disengaged.
c. Insufficient fluid in reservoir.
d. Faulty pressure or return line.
e. By-pass valve in rockshaft control valve sticking.
f. Faulty pump, paragraph 191.
g. Faulty rockshaft control linkage, paragraph 196.
h. Remote cylinder control levers not in neutral position.
i. Rockshaft control valve gasket or "O" ring has failed, paragraph 200.
j. Restricted oil return passages.

177E. **HITCH SETTLES UNDER LOAD OR "HUNTS" IN TRANSPORT POSITION.** To determine whether the trouble is in the control valve housing or in the rockshaft piston, cylinder or associated parts, refer to paragraph

181A. The following are possible causes:

a. Rockshaft cylinder seal has failed, paragraph 199.

b. Rockshaft piston seal has failed, paragraph 199.

c. Check valve in rockshaft control valve housing is leaking, paragraph 184.

d. Check valve plug gasket leaking.

e. Rockshaft control valve gasket or "O" ring has failed, paragraph 200.

f. Rockshaft overload relief valve is leaking, paragraph 183.

g. Throttle valve gaskets leaking.

h. Porous or failed rockshaft cylinder or rockshaft housing.

i. Check valve improperly adjusted, paragraph 184.

177F. LESS THAN FULL ROCKSHAFT ROTATION. Could be caused by:

a. Control lever linkage too short, adjust linkage for 75 degree rotation.

b. Twisted load control yoke shaft.

c. Sheared load control arm key.

d. Load control springs are short, paragraph 188.

e. Loose or bent linkage.

f. Control valve spring unhooked or failed.

g. Insufficient oil in rockshaft housing.

177G. WORKING DEPTH CHANGES OR CONTROL LEVER MOVES DURING OPERATION. Could be caused by:

a. Friction brake adjustment loose on control shaft, paragraph 186.

b. Retaining ring upset from groove in control shaft, paragraph 196.

177H. HITCH FAILS TO LOWER OR LOWERS SLOWLY. Could be caused by:

a. Throttle valve closed, paragraph 179.

b. Weak or failed rockshaft return spring.

c. Failure in linkage.

d. Oil viscosity too high, paragraph 176.

e. Check valve stem lock nut loose, paragraph 184.

177I. RELIEF VALVE OPERATES CONTINUOUSLY. Could be caused by:

a. Control lever linkage too long, adjust rockshaft travel to 75 degrees.

b. Remote cylinder control lever not in neutral position.

c. Load too heavy.

d. Relief valve opening pressure too low, paragraph 182, 189 or 190.

177J. OIL OVERHEATS (LOAD AND DEPTH CONTROL SYSTEM). Could be caused by:

a. Relief valve operates continuously, paragraph 177R.

b. Insufficient oil in reservoir.

c. Excessive load response, paragraph 177B.

d. Excessive load.

e. Continued use of lift, stop once in a while to allow oil to cool.

f. Worn pump, paragraph 191.

177K. REMOTE CYLINDER WILL NOT LIFT LOAD OR WILL NOT FUNCTION WHEN NOT LOADED. Could be caused by:

a. Air in remote cylinder, paragraph 176.

b. Pump disengaged.

c. Faulty pump, paragraph 191.

d. System overloaded.

e. Relief valve opening pressure too low, paragraph 189 or 190.

f. Gasket or "O" ring failed between remote cylinder control valve housing and rockshaft housing.

g. Faulty oil lines, packings or couplings.

h. Porous control valve housing.

i. Control shaft Woodruff key failed or missing.

j. Missing inner check valve ball on return side, paragraph 193 or 194.

k. Single valve housing, or No. 1 circuit of dual valve housing set for single-acting cylinder operation, and hoses crossed, paragraph 178.

l. Thermal relief valve missing.

177L. REMOTE CYLINDER WILL NOT LOWER. Could be caused by:

a. Air in remote cylinder, paragraph 176.

b. Missing inner check valve ball on return side, paragraph 193 or 194.

c. Control shaft Woodruff key failed or missing.

d. Faulty oil line coupler or coupler not completely engaged.

177M. REMOTE CYLINDER CONTROL LEVER DOES NOT RETURN TO NEUTRAL. Could be caused by:

a. Failed centering spring, paragraph 193 or 194.

b. Dual valve housing assembled without secondary relief valve balls.

c. Valve housing porous between relief valve and detent passages.

d. Control valve snap ring failed or disengaged, paragraph 193 or 194.

e. Control valve sticking, paragraph 193 or 194.

f. Detent sticking, paragraph 193 or 194.

g. Single valve housing, or No. 1 circuit of dual valve housing set for single-acting cylinder operation, paragraph 178.

h. Single-acting cylinder by pass screw not seated in housing. Screw must be firmly seated. If early models have thick bodied by-pass screw, discard it and install the late thin-bodied screw.

177N. REMOTE CYLINDER CONTROL LEVER WILL NOT LATCH IN FAST OPERATING POSITION. Could be caused by:

a. Remote cylinder overloaded.

b. Relief valve pressure set too low, paragraph 189 or 190.

c. Detent stuck in the disengaged position, paragraph 193 or 194.

d. Detent spring has failed, paragraph 193 or 194.

e. Detent has failed, paragraph 193 or 194.

f. Defective oil line coupler or coupler not engaged.

177P. REMOTE CYLINDER SETTLES UNDER LOAD. Could be caused by:

a. Dirty check valve, paragraph 193 or 194.

b. Remote cylinder adapter packing has failed.

c. Single-acting cylinder by-pass screw open while using double acting cylinder, paragraph 178.

d. Valve housing assembled without outer check valve, paragraph 193 or 194.

e. Thermal or overload relief valve leaking, paragraph 193 or 194.

f. Remote cylinder rod seal leaking, paragraph 203.

g. Remote cylinder piston ring failure, paragraph 203.

h. Remote cylinder casting failure, paragraph 203.

i. Outer check valve ball leaking or missing, paragraph 193 or 194.

177Q. "POWR-TROL" OIL OVERHEATS (REMOTE CYLINDER SYSTEM). Could be caused by:

a. Control lever being held in engaged position after remote cylinder reaches end of stroke.

b. Low oil supply, paragraph 176.

c. Relief valve pressure set too high, paragraph 189 or 190.

d. Control lever does not return to neutral, paragraph 177M.

e. Rockshaft against stop at full raise. Adjust rockshaft travel to 75 degrees from lowered position.

177R. RELIEF VALVE OPERATES CONTINUOUSLY. Could be caused by control lever not returning to neutral, paragraph 177M.

177S. NOISY PUMP. Could be caused by:

a. Oil viscosity too high, paragraph 176.

b. Insufficient oil supply in reservoir, paragraph 176.

c. Air leak in oil return line.

d. Pump drive shaft oil seal leaking.

OPERATING ADJUSTMENTS

178. SINGLE OR DOUBLE ACTING CYLINDER ADJUSTMENT. To operate a single-acting cylinder on tractors equipped with a single cylinder valve housing, remove cap nut (Fig. JD1925) and slotted adjusting screw located under the nut. Thread the adjusting screw into the cap nut as far as it will go, then reinstall the cap nut on the valve housing. To operate a double-acting cylinder on tractors equipped with a single cylinder valve housing, remove the cap nut (Fig. JD1925) and turn the slotted adjusting screw into the valve housing until it seats; then reinstall the cap nut.

178A. On tractors equipped with a dual cylinder valve housing, adjustment of the No. 1 circuit for use with a single or double-acting remote cylinder is made the same as outlined in paragraph 178 except the cap nut and slotted screw are located on top of the adapter which attaches to right front of valve housing as shown in Fig. JD1926.

179. SPEED OF ROCKSHAFT DROP. The speed at which the rockshaft (and attached implement) drops is regulated by a throttle valve on side of rockshaft housing. To make the adjustment, remove cap nut (Fig. JD1927), loosen the jam nut and turn the slotted screw **in** to decrease or **out** to increase speed of rockshaft drop. Adjustment should provide a smooth drop of the implement. After adjustment is complete, tighten the jam nut

and install cap nut firmly on gasket seat.

180. ADJUST HYDRAULIC STOP REMOTE CYLINDER. Any length of stroke within an 8-inch range may be selected. To adjust the stroke, lift the piston rod stop lever, slide the adjustable stop along piston rod to the desired position and press the stop lever down. If clamp does not hold securely, lift stop lever, rotate it ½-turn clockwise and press in place. Make certain, however, that the adjustable stop is located on the piston rod so that stop arm contacts one of the flanges on stop.

181. LOAD CONTROL ADJUSTMENT. As shown in Fig. JD1928, the load control yoke has four attaching holes for the upper link. These holes determine the sensitivity of the load and depth control system. The bottom hole (No. 4) is the most sensitive and the top hole (No. 1) the least sensitive. The proper hole to use is usually specified in the John Deere implement operators' manuals. Also, markings under a pointer attached to the control yoke make it possible to see that the proper hole is being used.

The pointer must be adjusted properly, as follows: When there is no load on the control yoke, the rear edge of the pointer should align with the front edge of the front indicating mark. If it does not align, loosen the two nuts and reposition the pointer until proper alignment is obtained.

When tractor and implement are in operation, the rear edge of the pointer should float over the center mark below the pointer. If pointer is over front mark, there is not enough compression on the load control spring and the link should be moved to a lower hole. If the rear edge of the pointer is over the rear mark, there is too much compression on the load con-

trol spring and the link should be moved to a higher hole.

Note: On self-gauging implements for which load and depth control is not required, install the upper link in top (No. 1) hole in control yoke. When using self-gauging implements, the control lever should be pushed to the forward end of the quadrant.

SERVICE TESTS AND ADJUSTMENTS

181A. If the rockshaft will not stay up under load, the difficulty is in either the control valve housing or the rockshaft piston, cylinder or associated parts. To determine which unit is at fault, a test fixture (shown in Fig. JD1928A) is available from R.B. Precision Tools, 313 Morse St., Ionia, Michigan. Method of using the test fixture is as follows: With engine not running and implement lowered to the ground, install the test fixture in place of the regular throttle valve shown in Fig. JD1927. Make certain that the plunger is turned out partially. Start engine and raise implement all the way up, turn the test fixture screw all the way in until the valve seats

Fig. JD1927 — Location of throttle valve screw cap nut. The throttle valve controls the speed of rockshaft drop.

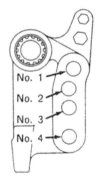

Fig. JD1928—Identification of holes in the load control yoke.

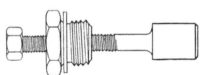

Fig. JD1928A—The test fixture shown can be used to determine the cause of the lift arms falling under a load. See text.

Fig. JD1925 — Single or double acting remote cylinder adjustment for models with a single cylinder valve housing.

Fig. JD1926—Single or double acting remote cylinder adjustment for models with a dual cylinder valve housing.

on rockshaft housing, then stop engine. If the implement lowers, the trouble is in the rockshaft piston, cylinder or associated parts. If the implement stays up, the trouble is in the control valve housing.

182. ROCKSHAFT RELIEF VALVE. When tractor is not equipped with remote cylinder valve housings, the rockshaft relief valve is located in a housing which is bolted to rear face of rockshaft housing.

To check and/or adjust the rockshaft relief valve opening pressure, remove the rockshaft throttle valve assembly from right side of rockshaft housing and connect a suitable pressure gage (at least 1800 psi capacity) to the opening from which the throttle valve was removed. Refer to Fig. JD1929.

Disconnect rod from the rockshaft control lever and with engine running and pump engaged, pull back on the rockshaft control arm to raise the rockshaft. The rockshaft will travel beyond its normal range and the rockshaft arm will strike the housing cover, preventing further travel of the rockshaft and causing the relief valve to open. The relief valve opening pressure as shown on the gage should be 1300-1400 psi. CAUTION: Operate system with relief valve open only long enough to observe the opening pressure.

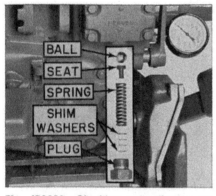

Fig. JD1929—Checking and adjusting the rockshaft relief valve opening pressure. Specified pressure is 1300-1400 psi.

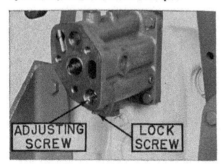

Fig. JD1930—Rockshaft cylinder overload relief valve adjusting screw. Specified relief valve opening pressure is 1450-1550 psi.

If the pressure is not as specified, remove the relief valve plug (Fig. JD1929) and add adjusting washers to increase the pressure or remove washers to decrease pressure. One washer changes the pressure approximately 40 psi.

When adjustment is complete, connect the rockshaft control rod, install the throttle valve assembly and adjust the rockshaft dropping speed as in paragraph 179.

When tractors are equipped with remote cylinder valve housing, the remote cylinder relief valves also protect the rockshaft circuit. Refer to paragraph 189 or 190.

183. ROCKSHAFT CYLINDER OVERLOAD RELIEF VALVE. The rockshaft cylinder overload relief valve is located in the rockshaft control valve housing which is bolted to front face of the rockshaft housing. Whenever the control valve is in neutral, the circuit to the rockshaft cylinder is sealed off from the rockshaft relief valve by the check valve. The overload relief valve then protects the cylinder circuit under these conditions. For example, the need for this protection can be appreciated because there could be considerable shock loads transmitted to the rockshaft cylinder and piston when transporting heavy implements over rough ground.

The overload relief valve is adjusted to open at 1450-1550 psi and should not normally require field adjustment. To check and/or adjust the overload relief valve opening pressure, proceed as follows:

Remove the rockshaft throttle valve assembly from right side of rockshaft housing and connect a suitable pressure gage (at least 1800 psi capacity) to the opening from which the throttle valve was removed. Refer to Fig. JD1929. Remove the rockshaft relief valve plug, adjusting washers and spring as shown. Replace the spring with a spacer made from a short piece of shaft, then screw the plug tightly

Fig. JD1931 — Exploded view of the rockshaft control valve housing, calling out the components to be concerned with in adjusting the check valve.

against the spacer to lock the relief valve ball against its seat.

Disconnect rod from the rockshaft control lever and with engine running and pump engaged, pull back on the rockshaft control arm to raise the rockshaft. The rockshaft will travel beyond its normal range and the rockshaft arm will strike the housing cover, preventing further travel of the rockshaft and causing the overload relief valve to open. The overload relief valve opening pressure as shown on the gage should be 1450-1550 psi.

CAUTION: Operate system with relief valve open only long enough to observe the opening pressure.

If the pressure is not as specified, remove the rockshaft (basic) housing assembly as outlined in paragraph 195. Remove the rockshaft control valve cover to expose the adjusting screw as shown in Fig. JD1930. Remove the lock screw and turn the adjusting screw **in** to increase or **out** to decrease the opening pressure. Each ½-turn of the adjusting screw will change the opening pressure approximately 50 psi. When adjustment is complete, install the lock screw and control valve cover.

Note: While the rockshaft housing assembly is removed, it is good practice to check the adjustment of the rockshaft check valve as in paragraph 184.

Install rockshaft housing assembly on tractor and connect the rockshaft control rod. Remove the relief valve plug (Fig. JD1929) and the home made spacer. Reassemble the relief valve as shown and be sure to install the same number of adjusting washers as were removed. Install the throttle valve assembly and adjust the rockshaft dropping speed as in paragraph 179.

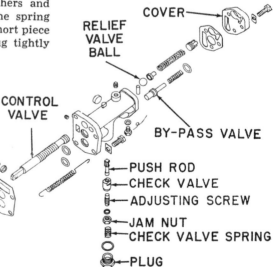

COVER
RELIEF VALVE BALL
CONTROL VALVE
BY-PASS VALVE
PUSH ROD
CHECK VALVE
ADJUSTING SCREW
JAM NUT
CHECK VALVE SPRING
PLUG

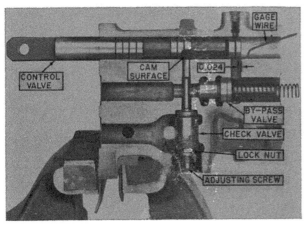

Fig. JD1932 — Cut-away view of the rockshaft control valve housing showing the points for checking the check valve adjustment.

184. ROCKSHAFT CHECK VALVE. To check and/or adjust the rockshaft check valve, remove the rockshaft control valve housing as outlined in paragraph 200. Remove cover (Fig. JD1931) from control valve housing. Also refer to Fig. JD1932 as well as JD1931. Remove check valve plug, check valve spring and check valve assembly. Position the control valve so the forward section covers about half of the passage which leads down to the by-pass valve. Reinstall the check valve and insert a home made spacer instead of the check valve spring. Then tighten the plug nut against the spacer so the check valve is firmly seated.

It should now be possible to easily move the control valve back and forth, a short distance, in the housing. If binding exists, the check valve is not seating because the check valve push rod is adjusted too long; in which case, remove the plug and spacer, loosen jam nut and back-off the adjusting screw enough to permit free movement of the control valve when the check valve is seated.

With the check valve seated, pull out on the control valve until the cam surface on the control valve contacts the check valve push rod; at which time, it should be possible to just insert a 0.024 diameter gage wire into the passage which leads down to the by-pass valve as shown in Fig. JD1932. If wire does not enter passage, remove the check valve and back-off the adjusting screw as necessary. If wire enters too freely, turn the adjusting screw in to decrease

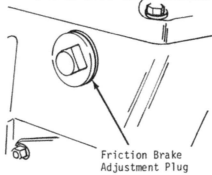

the clearance. Tighten the jam nut and recheck the adjustment as follows:

Remove the check valve plug and spacer. Insert gage wire in by-pass passage and push the control valve against the wire. Hold the check valve on its seat with your fingers. If the valve is properly adjusted, the slightest pulling of control valve out of the housing will start to open the check valve.

Install the check valve spring, retainer plug and housing cover. Install control valve housing assembly and rockshaft housing.

185. ROCKSHAFT ROTATION. The total rotation of the rockshaft when hydraulically operated should be 75 degrees. To adjust the rotation, lower the rockshaft to the lowest position and with engine running and pump engaged, adjust the length of the control rod (Fig. JD1933) at yoke so that when control lever is moved to the extreme rear position on the quadrant, the rockshaft will rotate 75 degrees up from its lowest position.

186. CONTROL LEVER FRICTION BRAKE. The control lever is held at selected positions by a friction brake on the control shaft in rockshaft housing. The brake should be adjusted so that a force of 12-15 lbs. applied at end of lever, is required to move the lever. To adjust the brake, remove plug (Fig. JD1934), exposing a self-locking nut on the 520, 620 and 720 or a cap screw on the 530, 630 and 730. Turn the nut or cap screw either way as required to obtain the specified force to move the lever.

187. DASH POT RELIEF VALVE. The dash pot relief valve is located within the rockshaft housing and can be removed as shown in Fig. JD1935,

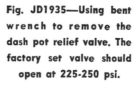

Fig. JD1934—Removal of plug will expose nut for adjusting the control lever friction brake.

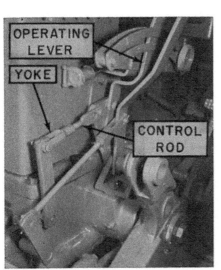

Fig. JD1933—Rockshaft operating lever, control rod and adjusting yoke. Full rotation of rockshaft should be 75 degrees.

Fig. JD1935—Using bent wrench to remove the dash pot relief valve. The factory set valve should open at 225-250 psi.

after removing the housing cover. The opening pressure of the valve is factory set to open at 225-250 psi and should not normally be adjusted. If the operation of the relief valve is questionable, it is recommended that a complete new valve unit be installed. If, however, a new valve is not available, remove the relief valve and attach it to a suitable hydraulic pump with a pressure gage. If the opening pressure is not as specified, turn the adjusting screw **in** to increase or **out** to decrease the pressure.

One method of checking the relief valve, using the hydraulic system of a tractor equipped for remote cylinder operation is shown in Fig. JD1936. Use a small steel ball and C-clamp to close off the bleed hole in the valve body as shown and be sure the shut-off valve is open. With engine running and hydraulic pump engaged, slowly move the control lever to the slow raise position and observe the pressure reading as the relief valve opens. NOTE: Be sure to hold a can over the valve to prevent undesirable spraying of oil.

188. LOAD CONTROL SPRINGS. The rear end of the load control springs should be carefully located with respect to the machined valve housing mounting surface on the rockshaft housing as shown in Fig. JD-1937. This dimension (D) can be checked after removing the load control yoke as in paragraph 197. Specified dimension (D) is as follows:

Series 520-530 ....1⅜ inches
Series 620-630 ....1$\frac{21}{32}$ inches
Series 720-730 ....1$\frac{7}{16}$ inches

The position of the springs is varied by adding or deducting shim washers in front of the springs.

189. SINGLE REMOTE CYLINDER VALVE HOUSING RELIEF VALVE. The relief valve in the single remote cylinder valve housing should open at 1230-1300 psi. To check the pressure, connect a pressure gage (at least 1800 psi capacity) and shut-off valve to the remote cylinder couplings as shown in Fig. JD1938. With the hydraulic pump engaged, start engine and allow to run for a few minutes before starting the test. With the shut-off valve open, move the control lever to the fast raise position. The control lever will remain in this position. Slowly close the shut-off valve and note the pressure gage reading the instant the control lever returns to neutral.

If the relief valve opening pressure is not 1230-1300 psi, remove the relief valve plug (Fig. JD1939) and add

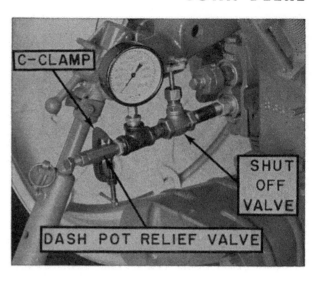

Fig. JD1936 — One method of testing the dash pot relief valve, using the hydraulic system of a tractor equipped for remote cylinder operation.

Fig. JD1937 — Using straight edge and scale to check position of the load control springs.

Fig. JD1938 — Pressure gage and shut-off valve hook-up for checking the single remote cylinder valve housing relief valve. This same hook up can be used to check the No. 1 circuit on dual valve housings.

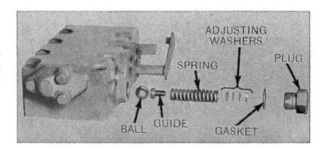

Fig. JD1939 — Pressure relief valve exploded from the single remote cylinder valve housing.

or deduct adjusting washers as required. Each washer will change the opening pressure approximately 40 psi.

190. DUAL REMOTE CYLINDER VALVE HOUSING RELIEF VALVE. The dual remote cylinder valve housing has two separate circuits, each having its own relief valve. The specified opening pressure is 1430-1500 psi for the number one circuit and 1230-1300 psi for the number two circuit.

To check the pressure for the No. 1 circuit, connect a pressure gage (at least 1800 psi capacity) and shut-off valve to the remote cylinder couplings as shown in Fig. JD1938. With the hydraulic pump engaged, start engine and allow to run for a few minutes before starting the test. With the shut-off valve open, move the control lever to the fast raise position. The control lever will remain in this position. Slowly close the shut-off valve and note the pressure gage reading the instant the control lever returns to neutral.

If the relief valve opening pressure is not 1430-1500 psi, remove the relief valve cap (Fig. JD1940) and add or deduct adjusting washers as required. Each washer will change the opening pressure approximately 40 psi.

To check the relief valve opening pressure for the No. 2 circuit, make certain that the control lever for the No. 1 circuit is in neutral position and connect a pressure gage (at least 1800 psi capacity) and shut-off valve to the remote cylinder outlets in the housing as shown in Fig. JD1940. With the hydraulic pump engaged, **start**

Fig. JD1941 — Exploded view of typical "Powr-Trol" pump.

3. Gasket
4. Expansion plug
5. Bushing
6. Pump housing
7. Dowel pin
8. Idler gear shaft pin
9. Gasket
10. "O" ring
11. Shifter
12. Shifter yoke
13. Washer
14. Thrust washer
15. "O" ring
16. Thrust washer
17. Bushings (Not sold separately)
18. Idler gear with bushings
19. Idler gear shaft
20. Oil seal
21. Drive gear
22. Bushing
23. Drive shaft
24. Woodruff key
25. "O" ring
26. Pump body
27. Pumping gears
28. Pump cover
31. Thrust washer
32. Idler shaft
33. Dowel pin
39. Packing

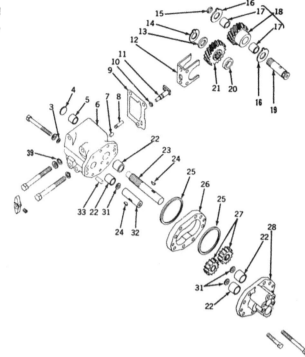

engine and with the shut-off valve open, move the control lever to the fast raise position. The control lever will remain in this position. Slowly close the shut-off valve and note the pressure gage reading the instant the control lever returns to neutral.

If the relief valve opening pressure is not 1230-1300 psi, remove the relief valve cap (Fig. JD1940) and add or deduct adjusting washers as required. Each washer will change the opening pressure approximately 40 psi.

Fig. JD1940 — One of the pressure relief valves exploded from the dual remote cylinder valve housing.

PUMP UNIT

191. R&R AND OVERHAUL. To remove the hydraulic pump, first drain the hydraulic system and disconnect the oil lines from pump cover. Withdraw the pump from governor housing and be careful not to lose the small pin (8—Fig. JD1941) which retains the idler gear shaft (19) in governor case. Also, notice the number of gaskets (9) used between pump and governor. Unbolt and remove pump cover; at which time, the need and procedure for further disassembly will be evident.

After pump is disassembled, wash all parts in solvent and thoroughly examine them for damage or wear. Inspect mating surfaces of housing (6), body (26) and cover (28) for wear or scoring. Inspect all bushings and renew any which show wear.

Pump specifications are as follows:

I.D. of pump body 2.452-2.454

O.D. of pumping gears . 2.4485-2.4495

Clearance between
  gears and body 0.0025-0.0055

Thickness of pump body
  (Except 620 orchard) . 0.7895-0.7905

Thickness of pump body
  (620 orchard) 1.0625-1.0635

Thickness of pumping gears
  (Except 620 orchard) . 0.7864-0.7870

Thickness of pumping gears
  (620 orchard) 1.0594-1.0600

Clearance between pump
  cover and gears 0.0025-0.0041

I.D. of pumping
gear bore 1.0005-1.0011
O.D. of pumping
gear shafts 0.9994-1.0000
O.D. of drive shaft
outer end 0.8082-0.8092
I.D. of drive shaft
outer bushing.......0.8112-0.8122
I.D. of other bushings..1.0025-1.0035

When reassembling, renew all "O" rings and seals and fill the space between the lips of seal (20) with gun grease. Tighten the cover retaining screws to a torque of 36 Ft.-Lbs. and check to be sure that pump turns freely. If not, loosen the screws and tap the cover and body and recheck.

ROCKSHAFT RELIEF VALVE HOUSING

When tractor is not equipped with remote cylinder valve housings, the rockshaft relief valve is located in a housing which is bolted to rear face of rockshaft housing.

192. R&R AND OVERHAUL. To remove the rockshaft relief valve housing, thoroughly clean the housing and surrounding area and drain the oil from rockshaft housing. Unbolt and remove the relief valve housing.

Remove plug (37—Fig. JD1942) and extract the remaining parts. Thoroughly clean all parts and examine them for damage or wear. If ball (32) or seat (31) are renewed, seat the new ball by tapping it with a brass drift and hammer.

When reassembling, be sure to install the same number of adjusting washers (35) as were removed and

after the unit is installed on tractor, check the opening pressure as outlined in paragraph 182.

SINGLE REMOTE CYLINDER VALVE HOUSING

193. R&R AND OVERHAUL. To remove the single remote cylinder valve housing, thoroughly clean the housing and surrounding area and drain the oil from rockshaft housing. Disconnect linkage from the control shaft arm, disconnect the oil lines from adapter, then unbolt and remove the housing.

To overhaul the removed unit, refer to Fig. JD1943 and proceed as follows: Remove the oil line adapter, nut (28), by-pass screw (27) and the detent

assembly (29, 30, 31 and 32). Remove the relief valve assembly (34, 35, 36, 37, 38 and 39) making certain none of the adjusting washers (37) are lost or damaged. Remove the two check valve retainers (25), then remove both check valve assemblies (21, 22, 23, 24 and 26) making certain all valve parts are removed and none are lost. Remove cover (1), then loosen jam nut (4) and set screw (3). Control shaft (18) can then be moved far enough to remove Woodruff key (17), after which shaft can be removed from valve housing. Unbolt and remove control valve cover (15) making certain washers (14) are not lost or damaged. Move control valve arm (6) to position shown

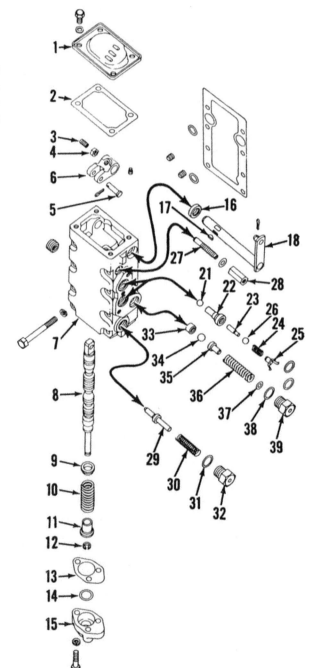

Fig. JD1943 — Exploded view of single remote cylinder valve housing typical of that used on all models.

1. Housing cover
2. Gasket
3. Set screw
4. Jam nut
5. Pin
6. Control valve arm
7. Valve housing
8. Control valve
9. Valve spring upper retainer
10. Control valve spring
11. Valve spring lower retainer
12. Snap ring
13. Gasket
14. Shim
15. Control valve cover
16. Oil seal
17. Woodruff key
18. Control arm and shaft
21. Check valve ball (2 used)
22. Check valve (2 used)
23. Check valve metering shaft (2 used)
24. Spring (2 used)
25. Retainer (2 used)
26. Check valve ball (2 used)
27. By-pass screw
28. Cap nut
29. Control valve detent
30. Detent spring
31. Washer
32. Plug
33. Relief valve seat
34. Relief valve ball
35. Relief valve spring guide
36. Relief valve spring
37. Washer
38. Washer
39. Plug

Fig. JD1942—Exploded view of the rockshaft relief valve and housing.

30. Gasket	35. Adjusting washers
31. Relief valve seat	36. Washer
32. Relief valve ball	37. Plug
33. Spring guide	39. Housing
34. Relief valve spring	40. "O" rings

in Fig. JD1944 and remove pin. The control valve assembly can now be removed from housing.

Inspect for cracks, porous conditions, excessive wear, missing or failed parts, dirt or metal particles and valves which do not seat properly.

The seat on outer check valves (22— Fig. JD1943) can be lapped in, using fine lapping compound. If ball seating surfaces on parts (22 and 33) or if steel balls (21, 26 and 34) are renewed, the ball can be seated by tapping it lightly against its seat with a brass drift and hammer. If oil seal (16) is renewed, install it with lip facing inward.

All parts should be lubricated with "Powr-Trol" oil before reassembly. Reinstall control valve assembly, installing pin (5) through control valve and control valve arm. Slide control shaft (18) through oil seal (16) and position Woodruff key; then move the control shaft into the control arm and secure with set screw (3) and jam nut (4). Install control valve cover (15) using the same number of washers (14) as were removed. Check free end play of control valve with dial indicator as shown in Fig. JD1945 and add or deduct washers (14—Fig. JD-1943) until the recommended free end play of 0.010 is obtained without compressing the neutralizing spring, then install cover (1).

Install both check valve assemblies (21, 22, 23, 24 and 26) and be

sure spring retainers (25) are secure. Install detent assembly (29, 30, 31 and 32) and by-pass screw (27). Install relief valve assembly (33, 34, 35, 36, 37, 38 and 39) using same number of adjusting washers (37) as were removed. Valve housing retaining bolts should be tightened to a torque of 36 Ft.-Lbs. After unit is reinstalled on tractor, check relief valve opening pressure as outlined in paragraph 189.

DUAL REMOTE CYLINDER VALVE HOUSING

194. R&R AND OVERHAUL. To remove the dual remote cylinder valve

housing, thoroughly clean the housing and surrounding area and drain the oil from rockshaft housing. Disconnect linkage from the control shaft arm, disconnect the oil lines from adapter, then unbolt and remove the housing.

To overhaul the removed unit, remove the adapter casting, refer to Fig. JD1946 and proceed as follows:

Remove retainers (79) and withdraw both of the No. 1 circuit check valve assemblies (74, 75, 76, 77 and 78). Unscrew plugs (57) and extract both of the No. 2 circuit check valves assemblies (51, 52, 53, 54, 55 and 56).

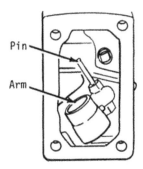

Fig. JD1944—Removing the control valve arm on single remote cylinder valve housing.

Fig. JD1945 — Checking control valve end play on single remote cylinder valve housing. Desired end play, without compressing the neutralizing spring, is 0.010.

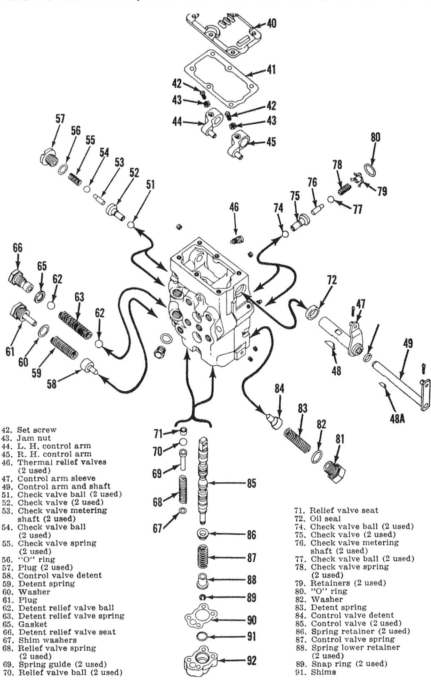

42. Set screw
43. Jam nut
44. L. H. control arm
45. R. H. control arm
46. Thermal relief valves (2 used)
47. Control arm sleeve
49. Control arm and shaft
51. Check valve ball (2 used)
52. Check valve (2 used)
53. Check valve metering shaft (2 used)
54. Check valve ball (2 used)
55. Check valve spring (2 used)
56. "O" ring
57. Plug (2 used)
58. Control valve detent
59. Detent spring
60. Washer
61. Plug
62. Detent relief valve ball
63. Detent relief valve spring
65. Gasket
66. Detent relief valve seat
67. Shim washers
68. Relief valve spring (2 used)
69. Spring guide (2 used)
70. Relief valve ball (2 used)

71. Relief valve seat
72. Oil seal
74. Check valve ball (2 used)
75. Check valve (2 used)
76. Check valve metering shaft (2 used)
77. Check valve ball (2 used)
78. Check valve spring (2 used)
79. Retainers (2 used)
80. "O" ring
82. Washer
83. Detent spring
84. Control valve detent
85. Control valve (2 used)
86. Spring retainer (2 used)
87. Control valve spring
88. Spring lower retainer (2 used)
89. Snap ring (2 used)
91. Shims

Fig. JD1946—Exploded view of the dual remote cylinder valve housing, typical of that used on all models.

Remove plug (81) and the No. 1 circut detent assembly (82, 83 and 84). Remove plug (61) and the No. 2 circuit detent assembly (58, 59 and 60). Unscrew plug (66) and the secondary relief valve assembly (62, 63 and 65). NOTE: Make certain none of the parts from any of these assemblies are mixed, lost or damaged. Remove cover (40), loosen nuts (43) and set screws (42), then slide control arm (47) enough to remove key (48). Withdraw both control shafts and control arms. Remove Woodruff key (48A) and separate the control arms. Unbolt and remove both control valve covers (92), then withdraw both primary relief valve assemblies (67, 68, 69 and 70). NOTE: Washers (67 and 91) must not be lost, damaged or mixed with washers from the other circuit. Withdraw control valves from bottom of housing.

Inspect for cracks, porous conditions, excessive wear, missing or failed parts, dirt or metal particles and valves which do not seat properly.

The seat on outer check valves (75 and 52) can be lapped in, using fine lapping compound. If ball seating surfaces on parts (52, 71 and 75) or if steel balls (51, 54, 70, 74 and 77) are renewed, the ball can be seated by tapped it lightly against its seat with a brass drift and hammer. If oil seal (72) is renewed, install with lip facing inward. Both thermal relief valves (46) should be renewed if their condition is questionable.

All parts should be lubricated with "Powr-Trol" oil before reassembly. Reinstall both control valve assemblies (85, 86, 87, 88 and 89) and both primary relief valves (68, 69 and 70). Install the same number of washers (67 and 91) and install control valve covers (92). NOTE: Make certain none of the washers (91 and/or 67) are lost, damaged or mixed with those of the other circuit. Reinstall control shafts and control arms (44 and 45), then tighten set screws (42) and jam nuts (43). Check free end play of each control valve with a dial indicator in a manner similar to that shown in Fig. JD1945 and add or deduct washers (91—Fig. JD1946) until the recommended free end play of 0.010 is obtained without compressing the neutralizing springs, then install cover (40). Reinstall all valve and detent assemblies in the sequence shown in Fig. JD1946. Valve housing retaining bolts should be tightened to a torque of 36 Ft.-Lbs. After unit is reinstalled on tractor, check each relief valve opening pressure as outlined in paragraph 190.

ROCKSHAFT HOUSING AND COMPONENTS

195. **R&R ASSEMBLY.** To remove the complete rockshaft (basic) housing assembly, first drain the hydraulic system and main case. Remove tractor seat and if three-point hitch is installed, remove the upper links, draft links and lift links. Disconnect the draft link supports from the drawbar and loosen the supports attached to rockshaft housing. Remove the platform and disconnect the hydraulic lines. Disconnect battery cable and wiring harness at rear of tractor. Support the complete housing assembly and attached units in a chain hoist so arranged that the complete assembly will not tip. Unbolt housing assembly from rear axle housing and move the complete assembly away from tractor.

CAUTION: This complete assembly is heavy and due to the weight con-centration at the top, extra care should be exercised when swinging the assembly in a hoist.

Reinstall the assembly by reversing the removal procedure and make certain that powershaft splines engage properly.

196. **CONTROL SHAFT AND LINKAGE.** To remove the control shaft and linkage, remove seat, batteries, battery box and rockshaft housing cover. Disconnect control rods. Remove access plug from left side of rockshaft housing and remove nut or cap screw (N—Fig. JD1948 or JD1949) and spring (S). Pull control shaft toward

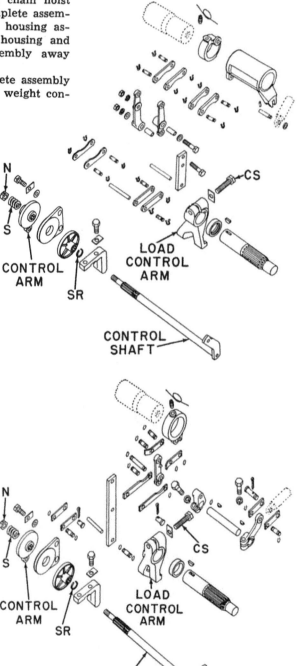

Fig. JD1948 — Exploded view of series 520 rockshaft control linkage, showing the relative location of all components. Series 530 is similar except a cap screw is used instead of nut (N).

Fig. JD1949 — Exploded view of series 720 rockshaft control linkage. Except for minor detailed differences, the series 620 linkage is similar. On series 630 & 730, a cap screw is used instead of nut (N).

right, disengage snap ring (SR) from shaft, then withdraw the shaft and snap ring. The need and procedure for further disassembly is evident.

The control plate lining must be in good condition and the aluminum control discs must not be scored, grooved or cracked.

When reassembling, be sure to securely seat snap ring (SR) in groove of control shaft. If early production tractors have a spring steel snap ring, discard it and install the late production soft steel ring. Be sure the short spline on the control shaft matches the corresponding groove in the control arm.

Adjust the control lever friction brake as in paragraph 186.

197. LOAD CONTROL YOKE. To remove the load control yoke, remove seat, batteries, battery box and rockshaft housing cover. Bend a ¾-inch wrench as shown in Fig. JD1950 and remove the dash pot relief valve as-

Fig. JD1952 — Rear view of 620 & 630 rockshaft housing, showing the load control yoke installation. Other models are similarly constructed.

Fig JD1953 — Using cap screw, washer and pry bar to remove load control shaft from rockshaft housing.

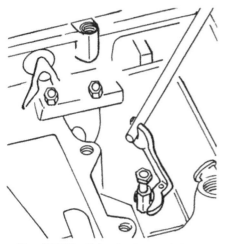

Fig. JD1950—Using bent wrench to remove the dash pot relief valve. The factory set valve should open at 225-250 psi.

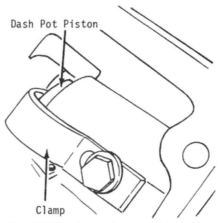

Fig. JD1954—Using the special tool shown in Fig. JD1951 to clamp the dash pot piston forward.

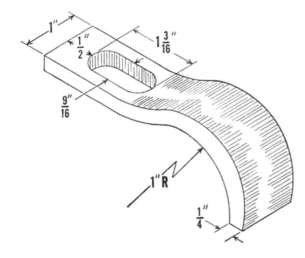

Fig. JD1951 — Special tool which can be made and used to clamp the dash pot piston forward.

sembly, but be careful not to disturb the adjusting screw which controls the relief valve opening pressure. Make up a tool, using the dimensions shown in Fig. JD1951, and using a ½ x 1-inch cap screw and washer, install the tool as shown in Fig. JD1954 and leave the screw slightly loose. Using a large screw driver or pry bar, pry the dash pot piston forward and tighten the clamp retaining screw securely. Loosen the load control arm retaining cap screw (CS—Fig. JD1948 or JD1949) and drive the roll pin (Fig. JD1952) out of the yoke. Remove the coupler bracket (or cover), screw a ⅜ x 4½-inch cap screw and washer into the load control shaft and using a pry bar as shown in Fig. JD1953, remove the load control shaft. Remove the load control yoke and withdraw the yoke roll pin from pocket in housing.

The need and procedure for further disassembly is evident. Before installing the load control yoke, check the control spring adjustment as outlined in paragraph 188.

Install the remaining parts by reversing the removal procedure.

198. ROCKSHAFT. To remove the rockshaft, remove seat, batteries, battery box and rockshaft housing cover. Remove lift arm from left end of rockshaft and disconnect the return spring. Remove lift arm from right end of rockshaft, then unwire and remove the set screw retaining the operating ring to the shaft. Refer to

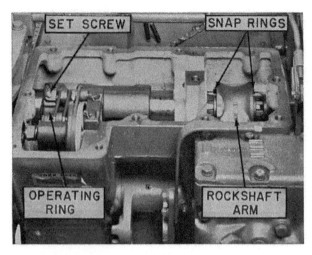

Fig. JD1955—Top view of rockshaft housing with top cover removed.

cylinder piston, first remove the rockshaft and rockshaft arm as outlined in paragraph 198. Refer to Fig. JD-1958, remove cap nut, loosen the jam nut and remove the throttle valve adjusting screw. Lift the rockshaft piston from the cylinder (Fig. JD1956).

To remove the cylinder after piston is out, first remove the control shaft as outlined in paragraph 196; then remove the cylinder retaining snap ring and lift cylinder from rockshaft housing. An L-shaped bar will be helpful in lifting out the cylinder.

Examine all parts and renew any which are damaged or worn. Renew

Fig. JD1955. Disengage the snap ring, (SR—Fig. JD1956) located just to the right of rockshaft arm (A), from its groove in the rockshaft and bump the rockshaft toward left until the snap ring (SR), located just to the left of the rockshaft arm, can be disengaged from its groove. Withdraw rockshaft from left and remove the rockshaft arm and piston rod assembly from above. Remove and discard the shaft oil seals.

Inspect the rockshaft bearing journals and bushings located in rockshaft housing. Replacement bushings are pre-sized and if not distorted during installation will require no final sizing. When installing the bushings, be sure to align oil hole in bushing with oil passages in the housing.

When reassembling, slide rockshaft through left side of housing, position the operating ring and on 520 and 530 only, slide the shaft through the control collar, making certain that slot in collar engages pin in rear end of control valve. Push shaft just through the center bearing and install one of the snap rings (SR) over end of shaft with long edge of ends of snap ring toward left. Position the rockshaft arm and piston rod assembly in housing and slide shaft just through the arm but make certain that "V" mark on shaft is in register with "V" mark on arm as shown in Fig. JD1957. Position right snap ring on shaft so that long edge of ends of snap ring is toward right. Push the rockshaft into position and engage the snap rings in the shaft grooves. Install the operating ring set screw and safety wire. Install the oil seals with lips facing inward. Install lift arms so that "V" mark on arms is in register with "V" mark on ends of shaft.

199. ROCKSHAFT OPERATING CYLINDER. To remove the rockshaft

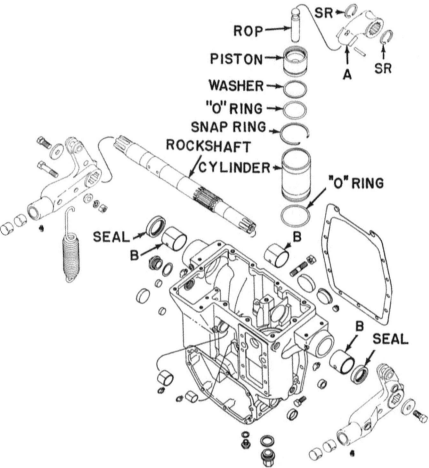

Fig. JD1956—Typical rockshaft housing showing an exploded view of the rockshaft, rockshaft operating cylinder and associated parts. Bushings (B) are pre-sized.

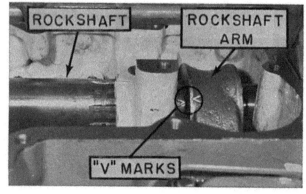

Fig. JD1957 — When assembling the rockshaft and rockshaft arm, the "V" marks must be in register.

piston rod if it is worn or scored at piston contacting end.

When reassembling, install "O" ring on cylinder, lubricate the "O" ring with gun grease and push cylinder into housing. Install the cylinder retaining snap ring with open end toward front of tractor. Install the piston "O" ring and backing washer with backing washer toward open end of piston. Lubricate piston with "Powr-Trol" oil and push piston into cylinder. Install rockshaft as in paragraph 198 and control shaft as in paragraph 196. Adjust the control lever friction brake as in paragraph 186 and the throttle valve as outlined in paragraph 179.

200. ROCKSHAFT CONTROL VALVE. To remove the rockshaft control valve, first remove the complete rockshaft (basic) housing assembly as outlined in paragraph 195. The control valve installation on front of rockshaft housing is shown in Fig. JD1959.

Note: Be sure to check for the possibility of leakage at control valve cover, check valve plug and between control valve housing and rockshaft housing. System malfunctions could be caused by leakage at these points and further disassembly would be unnecessary.

Remove cover from top of rockshaft housing and on 520 and 530, disconnect the control valve from the operating sleeve by unhooking the return spring, removing the spacer and pulling out the pin. On 620, 630, 720 and 730, disconnect the control valve from the operating linkage by unhooking the return spring, pulling the cotter pin and removing the connecting pin. On all models, unbolt and remove the control valve housing from rockshaft housing.

Fig. JD1959—Rockshaft control valve installation on front of rockshaft (basic) housing.

Remove cover from housing and remove pin (6—Fig. JD1960), return spring (26), control valve (1) and by-pass valve (13, 14 and 15).

Remove plug (22), spring (23) and check valve (18, 19, 20, 24 and 25). Remove cap screw (17), unscrew plug (10) counting the number of turns required to remove it so it can be re-

installed to the same position. Remove the overload relief valve (7, 8 and 9).

Check all parts for damage or wear and renew any which are questionable. Control valve (1) and by-pass valve (15) must slide freely in the housing bores. Examine the check valve and its seat in housing and check the seal with Prussian blue. Valve can be reseated by using fine lapping compound. The overload relief valve ball seat can be reconditioned by lapping with fine compound. To do so, make up a lapping tool by welding a 10-inch rod to a new relief valve ball (Part No. F 2772R). Be sure to use a new ball when reassembling. Install spring (9) and turn the nut (10) in the same number of turns that were required to remove the nut. Install cap screw (17). Examine finished surface near rear end of control valve (1) where the letter "T" is stamped. Install valve with letter "T" toward top of housing. Install by-pass valve (15) and spring (14). Install check valve and carefully adjust its position as outlined in paragraph 184. Install return spring and its pin (6). Position new "O" ring and gasket and install cover (11). Test the control valve housing assembly as in the following paragraph 201, then reinstall the control valve by reversing the removal procedure.

201. TESTING HOUSING - OFF TRACTOR. The overload relief valve can be tested, as outlined in paragraph 183, by installing the unit on the tractor; but, this entails considerable assembly and disassembly.

Fig. JD1960 — Exploded view of the rockshaft control valve housing and components. The overload relief valve (7, 8, 9 and 10) should open at 1450-1550 psi.

1. Control valve
2. Pipe plug
3. Valve housing
4. Plug
5. Pipe plug
6. Dowel pin for return spring
7. Relief valve ball
8. Spring guide
9. Relief valve spring
10. Relief valve adjusting screw
11. Cover
12. Gasket
13. "O" ring
14. By-pass valve spring
15. By-pass valve
16. Washer
17. Cap screw
18. Check valve push rod
19. Check valve adjusting screw
20. Jam nut
21. Washer
22. Plug
23. Check valve spring
24. Washer
25. Check valve
26. Return spring
27. Gasket
28. "O" ring

Fig. JD1958—Throttle valve screw cap nut and jam nut.

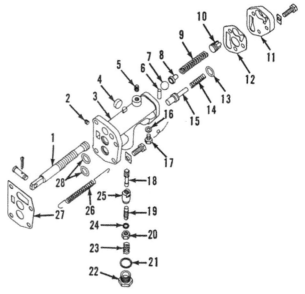

A means of testing these valves off the tractor is shown in Fig. JD1961, wherein the use of the pump on a hydraulic press is shown, but any hydraulic pump with a capacity of about 1800 psi can be used.

An adapter plate (Fig. JD1962) must be made and assembled to the control valve housing with "O" ring seals and a gasket.

Pump the pressure up to near the overload relief valve opening pressure of 1450-1550 psi and close the shut-off valve. The pressure may drop with the valve closed if the overload relief valve ball is not entirely seated. The ball will not seat completely until the film of oil is displaced from the seat. If this occurs, pump up the pressure two or three times to give the ball a chance to seat. If the seal is good, the pressure should then stay up. If it doesn't, either the overload relief valve or check valve is leaking.

Leakage past the rockshaft check valve will be indicated by the appearance of oil at hole (A). If the overload relief valve leaks, oil will appear at opening (B).

Be careful when building up pressure. If the overload relief valve opens (moves away from its seat), oil that escapes past the valve will also appear at opening (B). As oil is pumped into the control valve housing, the pressure increase will be indicated on the gage and when the opening pressure of the overload relief valve is reached (1450-1550 psi), the opening of the valve

Fig. JD1962—Home-made adapter plate used with the test layout shown in Fig. JD1961.

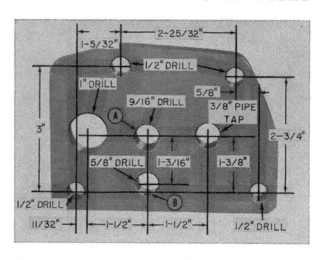

will be indicated by a slight drop in pressure.

If the overload relief valve does not open at the specified pressure, adjust the pressure by means of the adjusting nut (10—Fig. JD1960). Lock the adjustment by replacing the lock screw (17) so it engages a castellation on the nut.

If the valves leak, further lapping is required as outlined in paragraph 200.

202. **DASH POT PISTON.** To remove the dash pot piston, first remove the complete rockshaft (basic) housing assembly as outlined in paragraph 195. Compress the spring retainer with a tool similar to that shown in Fig. JD1964, remove snap ring (7—Fig. JD-1963) and extract the remaining parts.

Renew any damaged or worn parts and be sure the opening in the dash pot piston is open and clean. Lubricate all parts and reassemble

REMOTE CYLINDER (HYDRAULIC STOP TYPE)

203. **OVERHAUL.** To disassemble the unit, remove oil lines and end cap (11—Fig. JD1965). Remove stop valve (7) and bleed valve (5) by pushing stop rod (1) completely into cylinder. Withdraw stop valve from

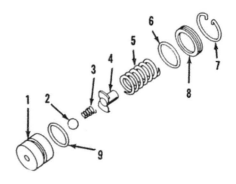

Fig. JD1963—Dash pot components.

1. Piston
2. Check valve ball
3. Check valve spring
4. Check valve cup
5. Piston spring
6. "O" ring
7. Snap ring
8. Spring retainer
9. "O" ring

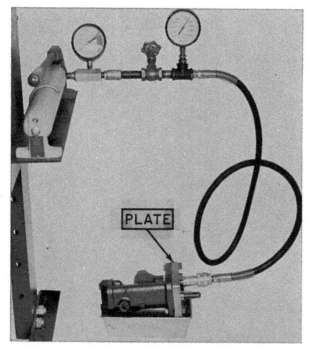

Fig. JD1961 — Using pump of hydraulic press, shut-off valve, gage and adapter plate shown in Fig. JD1962 to test the rockshaft control valve housing when unit is off tractor.

Fig. JD1964—Using a suitable home-made tool to compress the dash pot piston spring and retainer.

bleed valve, being careful not to lose the small ball (3). Remove nut from piston rod, being careful not to distort the rod and remove piston and rod. Push stop rod (1) all the way into cylinder and drift out Groov pin (19). Remove piston rod guide (27).

Examine all parts for being excessively worn and renew all seals. Wiper seal (30) should be installed with sealing lip toward outer end of bore. Install stop rod "V" seal assembly (20, 21, 22 and 26) with sealing edge toward cylinder. Complete the assembly by reversing the disassembly procedure and install the piston rod stop.

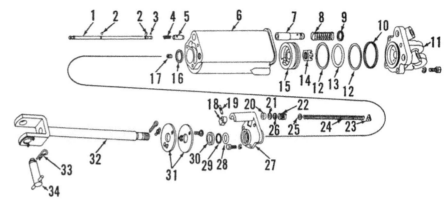

Fig. JD1965—Exploded view of hydraulic stop type remote control cylinder.

1. Stop rod	9. Gasket	18. Stop rod arm
2. Snap rings	10. Gasket	19. Groov pin
3. Ball	11. End cap	20. Packing adapter female
4. Bleed valve spring	12. Packing washer	21. "V" packing
5. Bleed valve	13. "O" ring	22. Packing spring
6. Cylinder	14. Nut	23. Stop rod washer
7. Stop valve	15. Piston	24. Stop rod spring
8. Stop valve spring	16. Piston rod guide gasket	25. Stop rod washer
	17. Pipe plug	26. Packing adapter male
		27. Piston rod guide
		28. "O" ring
		29. Packing washer
		30. Wiper seal
		31. Rod stops
		32. Piston rod
		33. Locking pin
		34. Attaching pin

APPENDIX 1
Pertaining to Standard Units Specifications and Adjustment Data

CARBURETOR

Marvel-Schebler Carburetor Calibration Data are as follows:

Model	Repair Kit	Gasket Kit	Float Valve & Seat Assembly	Idle Adj. Needle Assembly	Metering Adj. Valve Assembly	Nozzle	Power Jet
94	286-1123	16-671	233-543	43-665	43-661	47-385	49-188
95	286-1124	16-671	233-543	43-665	43-661	47-394	49-205
96	286-1222	16-671	233-543	43-665	43-661	47-412	49-192
97	286-1223	16-671	233-543	43-665	43-661	47-405	49-266
98	286-1224	16-671	233-543	43-665	43-661	47-405	49-347
99	286-1125	16-671	233-543	43-665	43-661	47-412	49-158
106	286-1123	16-697	233-543	43-665	43-661	47-385	49-188

Recommended float setting is ¾-inch from bottom of float to gasket surface.
Initial needle settings are: Idle Needles; seat lightly and open approximately 1½ turns.
　Power needles; seat lightly and open 2¼ turns.

DISTRIBUTORS

Delco-Remy Distributor Test Specifications are as follows:

1112569
Rotation, drive end..............CC
Contact gap 0.022
Cam angle range, degrees.......57-65
Distributor advance data in
　degrees at rpm:
　Start .... 0-2@250
　Intermediate .... 5-7@475
　Maximum ....9-11@650
1112576
Rotation, drive end..............CC
Contact gap 0.022
Cam angle range, degrees.......57-65

Distributor advance data in
degrees at rpm:
　Start ....0.5-3.5@375
　Intermediate ....4-7@450
　Maximum ....8.5-11.5@550

GENERATOR

Delco-Remy Generator Test Specifications are as follows:

1100309
Brush spring tension..........28 oz.
Field draw, volts................12.0
Field draw, amps...........1.58-1.67
Cold output, amps..............20.0
Cold output, volts..............14.0
Cold output, rpm..............2300

REGULATOR

Delco-Remy Regulator Test Specifications are as follows:

1118792
Ground polarity ....P

Cut-out relay
　Air gap 0.020
　Point gap 0.020
　Closing voltage, range.....11.8-14.0
　Closing voltage, adjust to......12.8

Voltage regulator
　Air gap 0.075
　Voltage setting, range.....13.6-14.5
　Voltage setting, adjust to.......14.0

STARTING MOTOR

Delco-Remy Starting Motor Test Specifications are as follows:

Motor Number	Volts	Brush Spring Tension Ounces	No Load Test			Lock Test		
			Volts	Amps	RPM	Volts	Amps	Torque Ft.-Lbs.
1107725	12	35	10.6	*112	3240	3.5	*385	..
1108155	12	24	11.3	70	6000	6.7	530	16
1113041	12	28-36	11.5	50	8000	3.25	500	22
1113079	12	48	11.5	50	6000	3.3	500	22
1113092	12	48	11.5	50	6000	3.3	500	22
1113093	12	48	11.5	50	6000	3.3	500	22
1113304	12	36-40	11.4	65	6000	5.0	725	44

*Includes Solenoid.

SOLENOID

Delco-Remy Solenoid Switch Test Specifications are as follows:

1119778 and 1119789

At 80 degrees F., consumption should be:

Both windings
Amps 72.0-76.0
Volts 10.0

Hold-in winding
Amps 18.0-20.0
Volts 10.0

STARTING MOTOR PINION ADJUSTMENT

Series 620-630-730

204. With starting motor and solenoid unit removed from tractor, mount same in a vise or holding fixture and disconnect the heavy copper strap between solenoid and motor at the starting motor terminal. Connect a wire from a 6-volt battery to a good ground on the motor, engage solenoid manually and hold solenoid in the engaged position by connecting another wire from the battery to the "S" terminal on the solenoid as shown in Fig. JD1967.

Hold the pinion (Fig. JD1968) toward the starting motor to remove any play in the linkage and move the arm-

ature as far as possible in the opposite direction. Now, using a feeler gage, check the clearance between end of pinion hub and inside edge of pinion retainer. If clearance is not 0.010-0.140, make any necessary adjustments by loosening the serrated coupling lock screw and moving the coupling link in or out as required. See inset in Fig. JD1967.

Series 530

205. The procedure for adjusting the starting motor pinion on series 530 is the same as outlined in paragraph 204, with one exception. In addition to connecting the battery as outlined, a jumper wire must be connected from the solenoid motor terminal to a good ground on starting motor as shown in Fig. JD1969.

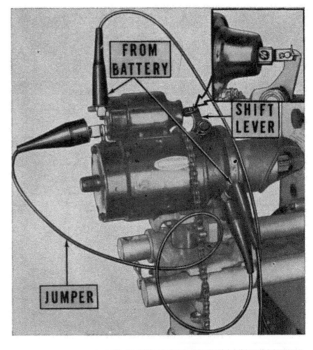

Fig. JD1969—Test hookup for checking the starting motor pinion adjustment on series 530.

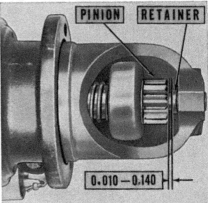

Fig. JD1967—Test hookup for checking the starting motor pinion adjustment on series 620, 630 and 730.

Fig. JD1968—Checking the starting motor pinion position on series 530, 620, 630 and 730.

NOTES

NOTES

NOTES

NOTES

NOTES

NOTES